The geometry of meaning

also by Arthur M. Young

The Reflexive Universe

The Bell Notes

Which Way Out? And Other Essays

Arthur M. Young

The geometry of meaning

ROBERT BRIGGS ASSOCIATES

Manufactured in the United States of America
Designed by Jerry Tillett

Second printing

Library of Congress Cataloging in Publication Data

Young, Arthur M. 1905–
The geometry of meaning.

Includes bibliographical references and index.
1. Knowledge, Theory of. 2. Meaning (Philosophy)
3. Whole and parts (Philosophy) I. Title.
BD161.Y68 121 76–918

ISBN 0–9609850–5–0

Robert Briggs Associates
P.O. Box 9
Mill Valley
CA, 94941

The Geometry of Meaning was
originally published by
Merloyd Lawrence

The publishers wish to thank Sheila La Farge
for extensive editorial assistance
in preparing this book
for publication.

The author wishes to thank Sam Brooks
for his interest and help in rewriting and simplifying
the first nine chapters of this book,
and Chris Bird, Alice Morris, Chris Clark
and Robert Temple
for editorial suggestions.

Contents

Introduction

"All meaning is an angle."

I don't know where I first encountered this enigmatic statement. I do recall that its origin was said to be in ancient Egypt, and I draw great comfort from this confirmation that there was at one time, perhaps so long ago that it was not even registered by Greek thought, a tradition that reflected the same conclusions I have reached after a lifelong effort to formulate meanings without reverting to the circularity found, for example, in dictionary definitions.

I would like to call this book an essay in philosophy. However, I should point out that I do not mean philosophy as it is usually practiced today, but in the older sense of "the science which investigates the facts and principles of reality." In this sense, philosophy not only encompasses the natural sciences, but explores the implications of the findings of science, and also deals with the relationship between the knower and the known.

Finding the abstruse mathematical formulations of science incomprehensible, modern philosophy falls back on the meaning of words. It therefore loses touch with the reality that gives science its strength and is the ultimate reference for the words upon which philosophy depends.

The ignominy of this retreat was recognized as early as

1905, when the American philosopher Charles Sanders Peirce berated metaphysicians for not pursuing truth more vigorously:

> "Its business (i.e., metaphysics) is to study the most general features of reality and real objects. But in its present condition it is, even more than the other branches of coenoscopy (the study of ordinary things) a puny, rickety, and scrofulous science. It is only too plain that those who pretend to cultivate it carry not the hearts of true men of science within their breasts. Instead of striving with might and main to find out what errors they might have fallen into, and exulting joyously at every such discovery, they are scared to look Truth in the face. They turn tail and flee her. Only a small number out of the great catalogue of problems which it is their business to solve have ever been taken up at all, and those few most feebly. Here let us set down almost at random a small specimen of the questions of metaphysics which press, not for hasty answers, but for industrious and solid investigation: Whether or not there be any real indefiniteness, or real possibility and impossibility? Whether or not there is any definite indeterminacy? Whether or not there be any strictly individual existence?"*

Though the list of questions continues for half a page, I have quoted only the first few as highly significant. In his question "Whether or not there is any definite indeterminacy?" Peirce anticipates Heisenberg's discovery of indeterminacy by more than twenty years. In asking "Whether or not there be any strictly individual existence?" he seems to sense the real flavor of quantum theory, with its discrete entities immersed in the continuum. Peirce's third question is thus equivalent to the continuing problem of the reconciliation of quantum theory (with its discrete particles) and relativity (with its continuum).

Peirce's questions, as he rightly observed, are metaphysical. They begin where science leaves off. The conflict between discreteness and continuity, so prominent in modern science, is

*_Philosophical Writings of Peirce_ (p. 314). Justus Buchler, ed. New York: Dover Publications, Inc., 1955.

only an old theological conflict in a new guise: whether individual freedom (like the quanta with their definite indeterminacy) is possible in a universe run by God (a continuum with implied uniformity).

Although Peirce was a philosopher, I am using this example to suggest that even the unspecialized thinker has access to basic questions which transcend technical proficiency; and that these basic questions are approachable insofar as one resists getting sidetracked by technical obfuscation, and maintains an awareness of the whole.

I take the position that a whole object or situation is divided into *aspects* (or, to use Aristotle's word, causes) and that these aspects have an angular relationship to one another. An account of this division is the purpose of this book. The ultimate goal is to regain the whole by knowing how the parts fit together. The separation of form from function, or fact from belief, has led to a strange and disconcerting hiatus in our present thinking. We have been so busy perfecting the means of travel that we have forgotten where we wanted to go.

Is my opening statement, "All meaning is an angle," too abstract? Not if one accepts my allegation that meaning is in general a kind of *relationship.* In terms of people, for example, let us take two men competing in a tennis match. They are *opponents.* They face each other. Their positions, both literally and metaphorically, are represented by an angle of 180 degrees, or a diameter. We refer to diameter when we say that two opinions are diametrically opposed.

A tennis match is judged by a referee who sits at one end of the net, at *right angles* to both opponents, surveying their interaction, but as an impartial witness, leaning neither way. The geometric or angular reference is not really so much difficult as it is disarming; it is so natural that we overlook it.

Some readers will find the geometry easy but will object to my interpretations, insisting that I put an improper "construction" on the technical devices of science. Yet not to transgress the presumed divisions between categories—not to attempt a translation of one into the other—would be to abandon the very purpose of my efforts, which is to devise a technique for structuring the components of meaning.

The customary procedure in any science is a process of induction from a myriad of particular observed phenomena to their description and explanation in terms of a few simple abstract properties. Since these fundamental terms lie deeper than definition, they are acknowledged as undefinable; and since they cannot be further reduced by the inductive method, it is presumed that they are truly basic. In this book, however, we postulate a *unity* whose division into aspects *creates* those undefined terms of science.

We find that simultaneous divisions of the whole into two, three, or four parts yields a meaningful description of these parts. Division into four is taken up first because it is the most easily grasped by rational thought. Next I will discuss the threefold division, and finally the twofold. Because it is so far beyond rational understanding, we can say little about the initial unity itself beyond describing it, in one sense, as the dynamic potency whose division creates a tension between the parts. In causing their interaction, this tension creates meaning.

Our method is technical only in that the manner in which the division of totality refines itself into a technique is similar to the manner in which measurement refined itself into geometry. Like this book, geometry came about through a quest for order. Geometry is an ordering so abstract that the roles of its elements, points and lines, can be interchanged, leaving its basic propositions intact: a line is determined by two points; a point is the intersection of two lines. The validity of geometry is independent of the *substantial* nature of the elements it employs. I believe equally fundamental relationships lie at the foundation of existence itself.

The geometry of meaning

I The fourfold division

Categories of knowing

What are the elements of an act of knowing?

At least two are immediately obvious: that which is known, and the consciousness which knows it. The knower's status is at least as important as that of the known, a point which modern physics has been forced to recognize in the uncertainty principle, which reveals the fact that, in order to know something about it, an observer must act upon, and thus disturb, an object. The act of knowing is thus a transaction between the object and the knower which involves *physical* exchanges of energy. Thus the concept of pure "parallelism"—a reflection within the knower of the known—is not tenable. We cannot think of the known as pure reflection within the knower, like the image of a flying swan on the smooth surface of a pond. When the swan lights on the water, its pure image is shattered by the broken surface.

Between observer and observed, then, there exists more than a simple duality. The complete description includes not only the actual physical impingements, the so-called sense data, of the known upon the observer, but also the projections of the observer upon the object. These projections are the source of error, and without them there would be no problem of correct knowledge. But they are not always erroneous. If I recognize a friend by the sound of his voice, I evoke an image of him that is substantially correct. We continually project properties on people and objects that we have learned to expect in them. It requires great mental effort to divest the world we experience of preconceptions and associations, if indeed it is possible at all.

The dynamic confrontation between knower and known, with the addition of the two kinds of relation between them—the objective information (sense data) coming from the object, and the nonobjective qualities *pro*jected by the knower upon the object—may be represented diagrammatically:

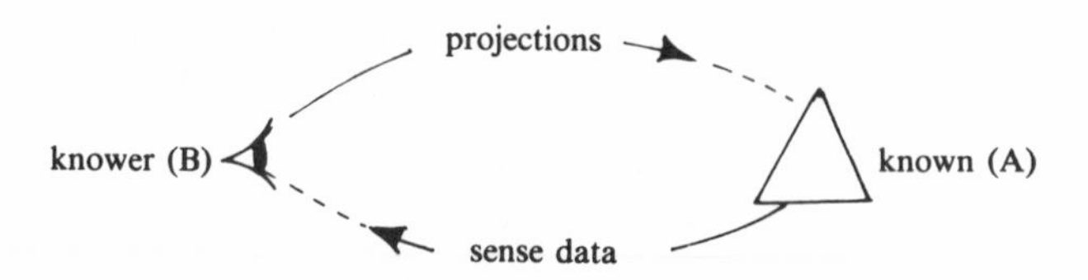

At least four kinds of relationship are involved:

1. AA—The relationships contained within the object itself: that it is, for example, an equilateral triangle, having three equal angles and sides. Such interrelationship provides the *definition* of the object. It holds for *all* equilateral triangles.
2. AB—The data which the knower receives from the object as sensations: its weight, color, texture, etc. This is *factual,* and concerns the particular object. It tells us that one particular triangle is dented, or broken, or that another triangle is blue.
3. BA—The qualities which the knower *projects* upon the object: that the triangle is beautiful or ugly, good or bad. This includes characteristics like solidity because such qualities are in part projected by the knower. Solidity is

not entirely objective since atomic structure is thought to be ninety-nine percent pure space.

4. BB—The *function* of the object for the knower: he uses the triangle for a watch fob or to illustrate an argument. This category consists essentially of relations of the knower to *himself* which he creates *for* the object, and would include his purpose in making it.

While there is no assurance that this approach will solve all problems, it has the merit of including all possible permutations of the relations between knower and object.

Only the first category (AA) is considered valid and useful in the scientific view. I propose that a complete description necessarily includes the three alternatives to this view.

The first category, that of the relationships contained within the object, conforms to the scientific requirement for a valid description; i.e., one to which all observers agree: a square has four sides, etc. The second category, sense perception of the object, while also objective and the basis for all scientific experiments, provides only a transition to scientific knowledge. It comprises the immediate data obtained from a particular object or experiment and, before it is applicable to new situations, must be formulated and generalized.

Hence, scientific knowledge is derived from observations, but is different from them. The observations consist of particulars, while scientific knowledge is general and belongs in the first category, not the second.

Thus we may describe the objectivity of the first category as *general,* that of the second category as *particular.*

The third category consists of values projected upon the object; for instance, its beauty or ugliness. When Hamlet says, "There is nothing good or bad but thinking makes it so," the implication is that value is not inherent in the object itself. This is the usual view, which I do not contest. But the further implication that value, because it is not inherent in the object, is *subjective,* or limited to the particular consciousness of a particular knower, deserves closer scrutiny. For the moment, I will simply point out that I am replacing "subjective" with the more accurate word "*pro*jective" for reasons that will shortly develop.

The fourth category, the function of the object, we also refer to as projective. An alternative key word is *orientation,* which may be interpreted both metaphorically and literally. It refers not only to the function of the object, which we described as a relation projected *for* the object rather than *upon* it, but also to the way in which the object relates to a larger context. It concerns "which way up" the carburetor is installed in the engine or the key is inserted in the lock. This aspect of the object is not objective. We cannot determine it by inspection of the object alone, but only by an awareness of the larger context of which the object is a part.

Such reasoning indicates the appropriateness of supplanting the word "subjective" by "projective." "Subjective" implies a personal description, with the possibility of illusion, while "projective" implies only that which is not objective, and leaves aside the question of reality.

Thus we say that an image is projected on a screen, or that the earnings of a corporation are projected for the year to come. The projective aspect can be of very practical significance. When Thiokol, a synthetic rubber intended for tires, was proposed as rocket fuel (changing its projective aspect), the price of stock in the company soared.

The distinction between general and particular, demonstrated for the first two categories, also applies to the third and fourth. We may call the third projective and general, the fourth projective and particular.

In the third category, valid projections, such as "Where

there's smoke there's fire," are possible only because such projection is general. The generality of projections also causes illusion. I see the beautiful tulips and imagine their smell and texture, only to discover that they are wax.

As for the fourth category, since function or purpose is an individual matter (my friend uses parking tickets to light the fire), the projection is particular. The word "orientation" implies this particularity; an orientation is by its very nature particular.

To illustrate the categories of knowing, let us use the example of an elephant:

1. The *formal description.* The structure of the elephant. The elephant as an object of scientific study. His anatomy, biology, behavior. Objective general.
2. The *sense data* which we experience by direct encounter with an elephant. The smell, the hairiness, the warm breath through the trunk. Objective particular.
3. The *values* we project on the elephant. How big he is! He seems kind, or patient, or terrifying, as the case may be. Projective general.
4. The *function.* The knower's interest in the elephant, what use he will make of him, as a circus attraction, a beast of burden, a zoological specimen, a source of meat or ivory. Projective particular.

The four aspects of a situation

In extending the method to deal with *situations* as well as objects, we must use more complicated examples. Suppose that a person is lost, but has a map. What kinds of information does he need to find his way?

1. The map itself. This, like the scientific description of the elephant, is an objective statement of the relationship of the distance between cities, of the location of roads, rivers, etc. The map is general and objective.
2. Where he is on the map. This information is different from the preceding. The person's position is particular as to place and time. It changes as he moves about. It is objective information but, unlike the map as a whole, it is *particular* and objective.
3. The scale of the map. Again, a different kind of information. It involves not just the scale in miles per inch on the map but the means of travel at the person's disposal. It is general and *projective.*
4. Orientation of the map. This different kind of information must be supplied by a compass. It also changes as the traveler moves about and is therefore particular. Note that it is an orientation, a direction, and that the direction of north and the direction in which the traveler wants to go are not necessarily the same. But in order to know in which direction to go, one must know how to orient the map. It is a relation of the map to the larger context. It is particular and projective.

In this example, there are again four aspects which correlate with four kinds of relationship:

Form correlates with the interrelationship of the parts of the known to one another.

Position correlates with the relationship of the traveler to points on the map.

Scale correlates with the relationship between a distance on the map and the effort the traveler is called upon to make.

Orientation correlates with the relationship of the traveler to a more ultimate reference which is not objective, his goal. This relationship is particular and projective.

Aristotle's causes

The concept of four basic categories of knowing, or four aspects of a situation, is not new—as is evident by reference to the famous four causes of Aristotle. Aristotle described an object (for example, a table) as having:

1. A *formal cause.* The blueprint or concept of the table, its shape and proportion. Corresponds to the objective general.
2. An *efficient cause.* The work of the carpenter in making the table. Corresponds to the objective particular. His particular work produced this particular table.
3. A *material cause.* The wood or other raw substance of which the table is made. The projective general. Wood is general because it can make many things besides tables. There is some ambiguity here since one might choose a particular piece of wood to make a table. However, both the work and the wood have general and particular aspects.
4. A *final cause.* The purpose of the table. Its function of holding things. The projective particular.

Whether we refer to them as causes, categories, or kinds of relationships between an observer and an object, we can recognize that the examples we have listed all involve an observer and an object or situation.

We need not expect exact correspondences between the different examples, but notice only that they are all subject to analysis in terms of the particular–general and objective–projective dichotomies, and demonstrate the sufficiency of a fourfold analysis.

The next step is to display the four categories graphically. If, in fact, we call the categories *aspects,* the graphic portrayal is almost mandatory, since the word "aspect" implies a direction from which something is viewed. The apparent sufficiency of exactly four categories suggests representing them as pairs of opposites on a cross axis. But before we attempt this, we must decide which categories should be considered opposites. For instance, should the formal cause be opposite the material cause, or the final cause? Other alternatives, such as in what direction to have each aspect face, or in what order to have the aspects, are arbitrary.

To decide which pairs are opposite, we shall refer to the two dichotomies of general versus particular, and projective versus objective. Since formal cause is objective and general, and final cause is projective and particular, they are doubly opposite and hence their relationship should be represented by the maximum angle, a full diameter of difference. Thus we have:

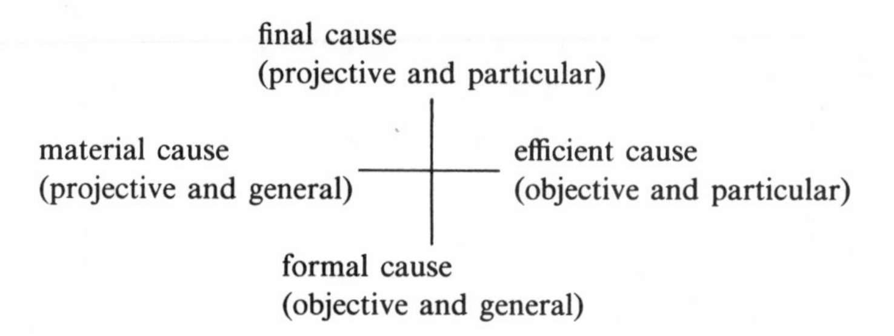

Several important properties of the four categories are made apparent by this graphic representation which, by correlating the aspects with directions in space, permits us to say:

1. Four aspects are sufficient for the analysis of a situation. If there were a fifth, it would be compounded of two others.
2. Four aspects are necessary to a situation. If, for example, we know of three, we should expect to be able to find a fourth.

3. Aspects can be formally related. Aspects 90 degrees apart are independent of one another. On the other hand, aspects at opposite ends of an axis are mutually opposed. This idea can be illustrated by reference to the four directions. We can move due east without moving north or south, but not without moving away from, or negating, west.

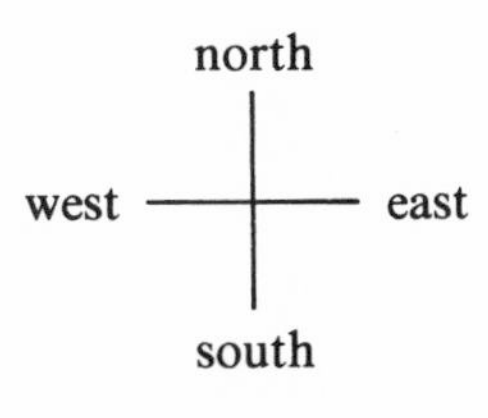

To illustrate the first point, we may refer again to the map. One could say, "In addition to the four aspects of the map already given, there is a fifth, the direction in which the traveler wants to go." But this direction is like a compass reading. It is an orientation, hence it is not a new category.

The second and third points will be illustrated later.

II The measure formulae

A mathematical method

The calculus, discovered simultaneously by Leibniz and Newton, is an important technical device which greatly enlarges the scope and power of the fourfold analysis. Since it is a foundation for the argument to come, I should describe it briefly.

The concept of the calculus consists of the recognition that, besides the measurement of space and time, we may also employ the *ratio* of space to time; that is, speed, or *velocity.* In these days of high-speed travel, it is difficult to understand why the discovery of a formal expression for such a household word as "velocity" raised such a fuss. But Bishop Berkeley, the philosopher, insisted that Newton's "fluxions," which were the ratio of an infinitesimally small increment of distance to an infinitesimally small increment of time, were "logical absurdities." "For," said the Bishop, "an infinitesimal is bad enough, but a ratio of infinitesimals is preposterous."

The person who insists he was not driving sixty miles an hour because he was traveling for only ten minutes is using the same reasoning.

Newton's great discovery was that we *can* speak of a velocity as having instantaneous value at a point, despite the fact that to measure velocity we have to take a finite distance and divide it by a finite interval of time.

Newton went further. "Not only may we speak of the rate of change of distance with time, which is *velocity,*" he said, "but we may speak of the rate of change of velocity with time, which is *acceleration.*" Velocity is expressed as dl/dt, which means the derivative (or ratio) of length (distance) with respect to time (dl signifies an arbitrarily small increment of length, dt an arbitrarily small increment of time); acceleration is expressed as $(d/dt)(dl/dt)$, the derivative of a derivative, or d^2l/dt^2. As position is measured by length (l), we will use the words interchangeably.

The calculus enabled Newton to solve the problem of planetary motion. Even more significant is the fact that the calculus, through its concept of derivatives, is able to give a precise definition to elusive notions like force, and hence makes possible the whole science of motion.

In following the use of the calculus to explore the four categories of knowing, the nonmathematically grounded reader should not permit a horror of formulae to spoil what can be an interesting adventure. Readers familiar with mathematics will hopefully be tolerant if their equipment is borrowed for a purpose that takes it beyond its usual employment.

To illustrate the interrelation of the concepts of the calculus —position, velocity, and acceleration—let us take the example of a swinging pendulum.

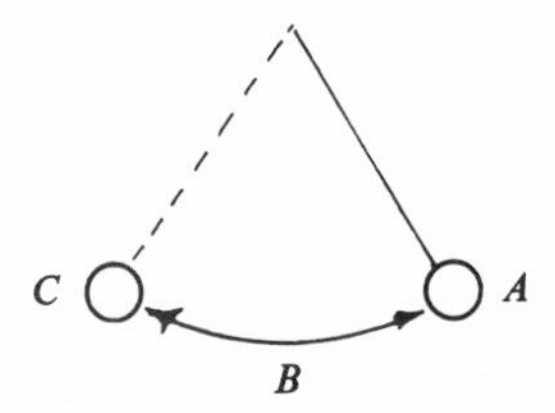

Assume that the bob of the pendulum is released at point A and swings from A through midpoint B to point C opposite A and back again. When it is to the right of midpoint B, its position is positive; when to the left, negative. *A* and *C* are thus its maximum positive and negative positions.

If we chart the position of the pendulum with respect to the passage of time, we may refer to the completion of one swing to the left and its return to *A* as a *cycle of action,* the halfway point of which is *C* and the first quarter of which is midpoint *B*. The third quarter will be *B* again but, to distinguish the midpoint of the return swing from that of the outgoing swing, the latter is called *D*. This can be represented by a new diagram on which the right and left positions of the pendulum are represented by a *vertical* distance above and below a horizontal line upon which *A, B, C, D, A′* represent successive time intervals.

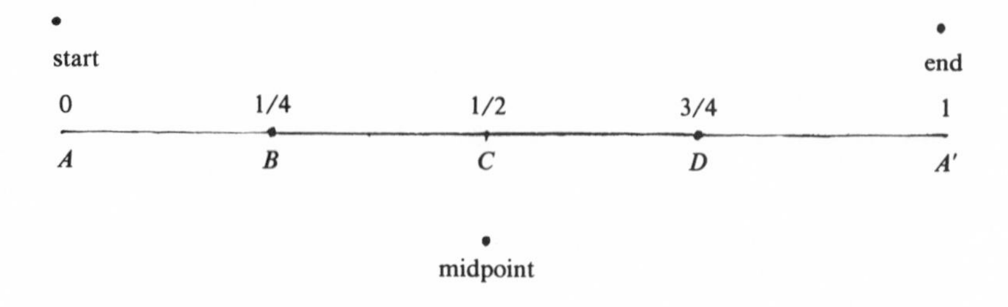

This vertical distance, or swing, will be maximum positive at *A,* zero at *B,* maximum negative at *C,* zero at *D,* and back to maximum positive at the end of the line, *A′*. A smooth curve is drawn through these points.

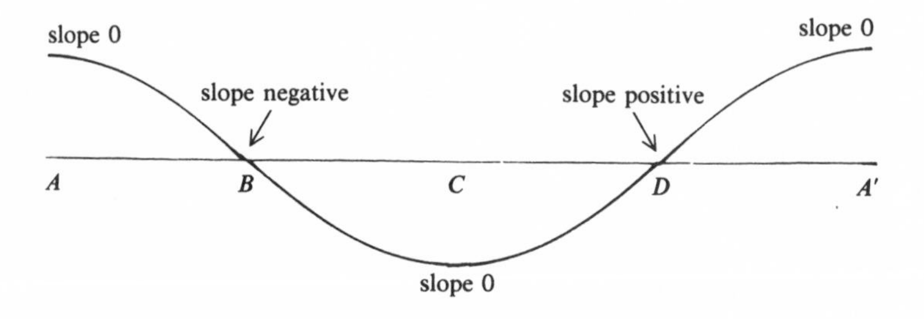

Note the slope, or "steepness," of this curve, which represents the *rate* at which the position changes.* At A it is zero (flat) because the position of the pendulum is changing very slowly; at the 1/4 point B, the slope is steep and downward (negative), at midpoint C it is zero again, at the 3/4 point D it is steep and upward (positive), and at the end it is again zero. These values are marked on the same diagram to obtain a new curve, which represents the value of the *slope* of the first curve.

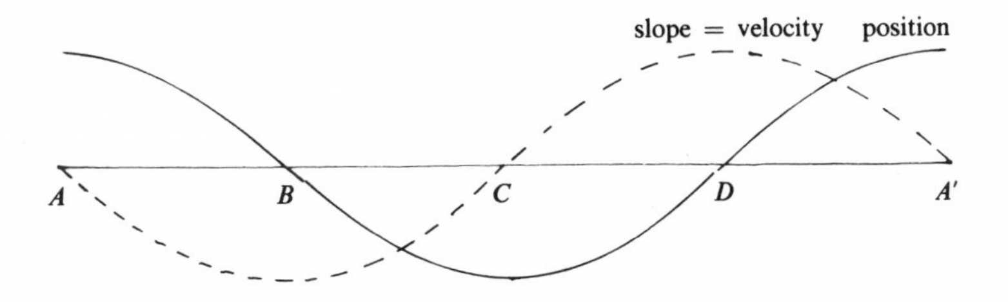

This second curve, showing the rate at which the position changes, depicts the *velocity* of the pendulum. Since position

*If we take a short section of the position curve, for example, where it first crosses the line, we can see that slope is the ratio of the vertical distance dl to the horizontal distance dt, and this ratio has a definite value.

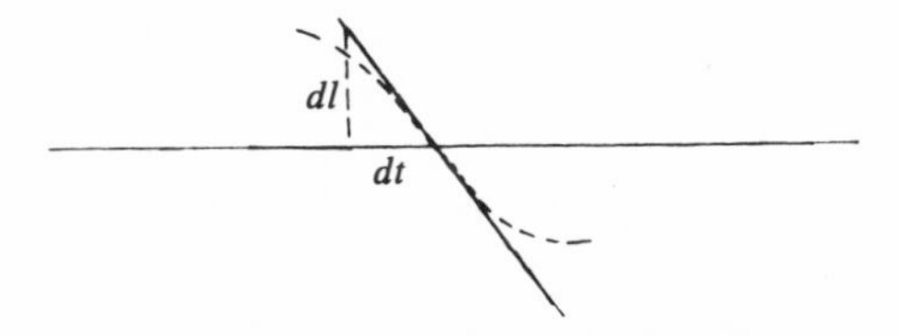

In the diagram we have made dl and dt large for visibility. But when dl and dt are large, the curve between them is not a straight line. However, if we make dl and dt small enough, the curve between them will approach a straight line, and the slope of the curve at point B will equal the ratio dl/dt, which is called the derivative of l with respect to t. Since l is distance and t is time, dl/dt is velocity. This was Newton's discovery. It permits us to deal with an entity that is not an immediate fact of sense experience. (Since an instantaneous picture of the pendulum would not reveal its velocity or even that it is moving, we have to take at least two pictures and plot their relationship.)

has been designated as positive when the pendulum is to the right, velocity is negative (moving to the left) in the first half of the swing. This is shown by the curve of velocity being below the line in the first half.

Going a step further, the slope of the velocity curve, which is to say the rate at which velocity is changing, is charted. This produces a third curve, which represents *acceleration.*

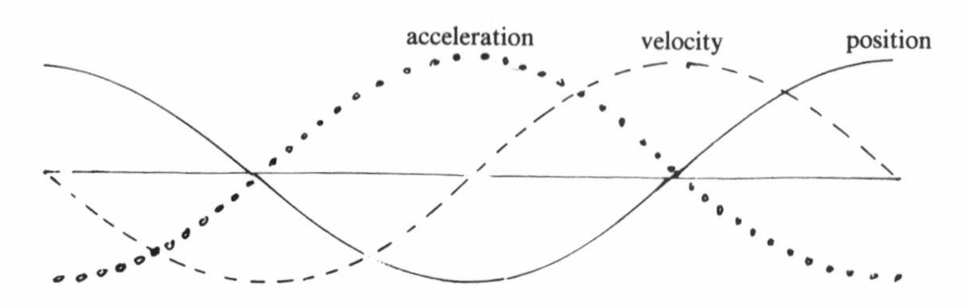

Observe that the acceleration curve is exactly opposite to the position curve. At the start, position is maximum and positive, acceleration maximum and negative. One may experience this by holding the pendulum at position *A* (positive), and feeling it pull *back* (negative). The pull is acceleration. If one lowers the pendulum into a vertical position, this pull reduces to zero (at *B*). The pull becomes maximum and positive (pulling toward *A*) when in position *C,* and so on.

Note that these curves are all of the same shape but the curve for velocity is displaced 1/4 of a cycle* *back* from the position curve, and the curve for acceleration is displaced 1/4 cycle back from the velocity curve and therefore 1/2 cycle from the curve of position. Acceleration is thus "out of phase" with position by 1/2 cycle. Velocity is out of phase with both acceleration and position by 1/4 cycle. Note that the peak of velocity comes at *D,* acceleration at *C,* and so on.

**A* to *A'* is a full cycle, so *A* to *B* is 1/4 cycle.

The cycle of action

To take a final step, the cycle of action is represented as a full circle. To do this, we simply bend the line *ABCDA′* around into a circle, with *A′* falling on *A,* and the points of maximum positive position, velocity, and acceleration falling on three points 90 degrees apart.

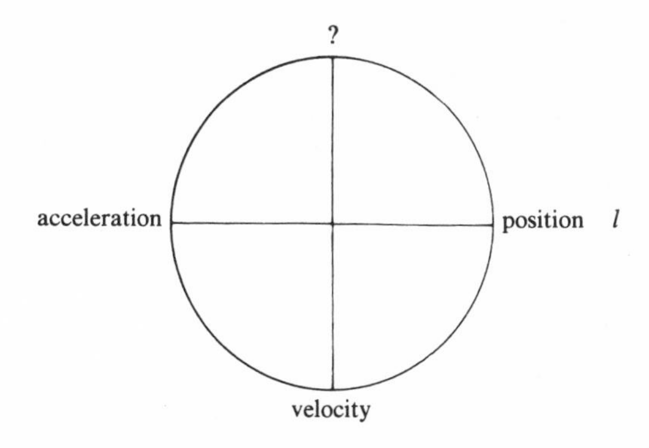

This charting presents the *phase relationship* between the three measures under discussion. Velocity lags behind position by 90 degrees, acceleration by 180 degrees, making acceleration opposite to position. The fourth point on the circle has no measure assigned to it. What could this measure be?

Since it is 90 degrees "beyond" acceleration, and each right angle has signified an additional derivative of position, we could expect it to be the third derivative, d^3l/dt^3. That is to say, the *rate of change of acceleration.* But what name and what meaning can we give to this?

The name given to the third derivative by aeronautical engineers is *jerk,* probably because when acceleration is changed by an automatic control, it does so in an all-or-nothing fashion which results in a jerk. But in the general case, as in human control, it need not be a jerk.

In any case, because this control factor is not described in the textbooks, we should give it special attention. Let us take the case of driving an automobile.

To increase the speed of an automobile, we push on the accelerator, causing positive acceleration. To decrease the speed, we step on the brake, causing negative acceleration. We may also alter the direction of the car by steering.* What is the process by which the accelerator, brakes, and steering change the acceleration of the car? Clearly, a change of acceleration is what we mean by the word *control,* which now goes in the diagram's fourth position.

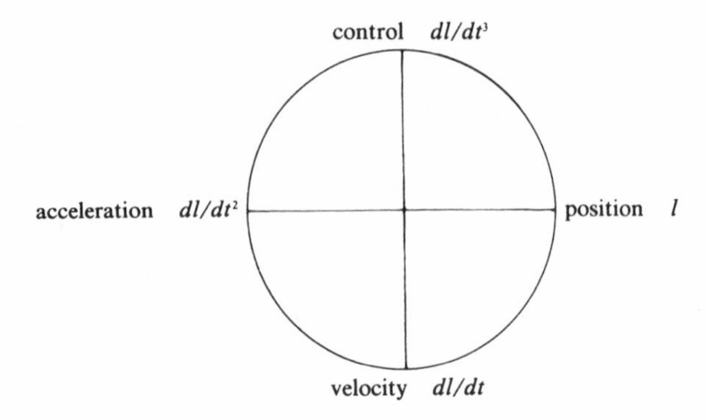

Reading clockwise, each measure is the change of the one before. This raises a question: what factor changes control?

*This is also a change of its acceleration, which may be understood as follows: Suppose we are traveling at a constant speed and make a U-turn without slowing down. We are now moving in the opposite direction, and our velocity, as measured from a point outside the vehicle, has changed from positive to negative. As this implies, a change of direction of velocity is, in fact, an acceleration. The force of this acceleration is what pushes objects in the car toward the outside of the curve. If we hold the wheel in one position so that the car travels in a circle, we are not changing the acceleration—we have simply imposed a constant acceleration.

What causes us to turn the automobile, or start it in the first place, or finally to stop it? The answer is "the destination." The destination is a place, or position. Thus the fifth step, the fourth derivative of position, that which determines control, *is the same as the first step.* This position is not the *same* position we started with but has the same category of measure (distance, or length), just as the direction in which the traveler wants to go is not a new category in the example of the map.

The examples of the map and the automobile both demonstrate the sufficiency of four categories. We need no fourth derivative. Position and its three derivatives are sufficient for analysis; and, I might add, necessary when there is control, human or otherwise.

It may be recalled that the problems of classical mechanics always exclude the possibility of a free agent "interfering with" the system. These problems employ only position and its two derivatives, velocity and acceleration, and in the case where friction can be ignored, may even omit velocity (for friction depends on velocity).

Thus the basic predictive equation that is used for treatment of the motion of all bodies, from atoms to spaceships, is:

$$d^2l/dt^2 + f(l) = K$$

In this d^2l/dt^2 = acceleration, $f(l)$ = an arbitrary function of length (position), K = constant.

The presumption in excluding a free agent is that there could be no prediction in such a case—the equation would be too complicated or would not apply.

However, the above argument shows that there *is* a formal expression which covers the "free agent," namely, the third derivative. This, of course, does not mean that prediction is possible. On the contrary, it means that freedom (or unpredictability) is part of the system. In certain cases, prediction is theoretically possible, as in the case of feedback wherein a predetermined position regulates control. One example of this is a target-seeking missile—but even here, although the control is not itself free, it has been set in advance by an agent who is free to choose the target.

One might suppose such a situation altogether too complicated for complete analysis, but at least it can be shown that the four categories are again sufficient:

1. To know the position of a body in space, we need *one* instantaneous observation (for instance, the photo finish of a race).
2. To know its velocity, which is computed from the difference in position of the body and the difference in time between the two observations, we need *two* such observations.
3. To know its acceleration, we need *three* observations.
4. To know that a body, for example, a vehicle, is under control, and thus distinguish it from one in which the controls are stuck, we need at least *four* observations. That is, we need three to know acceleration and one more to know that acceleration has been changed. (This still does not tell us the body's destination or goal.)
5. To know the destination, provided the operator does not change his mind or try to fool us, we need *five* observations.
6. To know the operator has changed his mind or is trying to fool us, we need *six* observations.

Note that the fifth observation is to establish a position (the destination) and the sixth a change of position. Thus categories five and six repeat the cycle, the fifth falling into the position category and the sixth into the velocity category. As in the case of the map, the sufficiency of four categories is demonstrated.

Use of the measure formulae

We have shown that the motion of a body, even one controlled by an operator, may be formally described by the four categories of measure. The form we have used to represent these measures is that employed in the calculus:

Position $= l$	Acceleration $= d^2l/dt^2$
Velocity $= dl/dt$	Control $= d^3l/dt^3$

There is an alternative, and simpler, representation in which the *d*'s are omitted and capital letters employed; thus:

Position $= L$	Acceleration $= L/T^2$
Velocity $= L/T$	Control $= L/T^3$

This is the form used in so-called *measure formulae,* the ten formulae used in physics to describe the motion of a body. Since I will discuss these formulae at some length, I will use the simpler representation.

In describing these measures—position and its derivatives—we have sought to establish:

1. That four categories of measure are necessary and sufficient for the analysis of motion of a moving body.
2. That the graphic representation of these measures as four right angles dividing the circle has a special significance—each right angle is a phase shift of 90 degrees and correlates with the derivative of the one before.

In the course of the above demonstration, we uncovered another point which has philosophical significance and suggests further implications:

3. That the measure technique of science can be extended to include free will, an aspect of a situation generally thought of as nonscientific.

What is going on? We have been standing on the sidelines watching the scientist ply his trade with some technical measurements that do not appear to concern philosophy; they are just the tools of his trade. But one of these, the third derivative, turns out to be the very thing discredited by science —the human, or free will, factor!*

Perhaps we should take a closer look at these measure formulae. Do they have a more general significance?

Categories of knowing represented by the measure formulae

What are these measure formulae, position and its derivatives, velocity and acceleration? Are they just the simple "physical quantities," as they are referred to in the textbooks?

Not so. Closer examination reveals fundamental *qualitative* differences between them. To understand this, consider how each of these measures is known. Position can only be *observed* visually or by less direct processes. Velocity is an intellectual abstraction: it cannot be known from direct experience. It must be computed. To know velocity, we must make two observations of position, determine their differences, and divide the time elapsed, thus obtaining a *ratio.* (Though velocity can be read from a speedometer, the only accurate speedometer, a chronometric tachometer, is itself a computer.) The velocity of the airplane even at six hundred miles an hour is something

*It has been suggested that rather than saying the third derivative is free will, I should say that the third derivative is where free will *enters*—but I would protest that this implies free will is something more than control.

that one is completely unaware of. The earth is "hurtling" through space at eighteen miles per second, yet we have no feeling of it.

Acceleration, however, is felt. For example, when an elevator suddenly starts down, you *feel* it in the pit of your stomach; when the airplane comes in for a landing and the jets are reversed, you are thrown forward in your seat. Acceleration may also be computed, but it can be directly and physically experienced by the knower because the nervous system registers change, not a steady state.

The notion of velocity is so far removed from direct human experience that it was not until after Newton's discoveries that it became a formally recognized concept. And for all its acceptance by modern civilizations, velocity is nevertheless denied by many philosophers, because it is neither an "event" nor an "individual."*

We may deny the reality if we wish, although it would then be hard to know which of the formulae we might endow with reality. In any case, this misses the point. What is important is that:

1. Position, velocity, and acceleration are separate and different aspects of the total situation.
2. They are all necessary.
3. They are included in the scientific description.
4. And most important, they are different categories of knowing. Position is observed, velocity is computed, acceleration is felt. Control, the fourth category, is essentially indeterminate.

*For an example of the confusion that exists over this question, see Smart, J. J. C., *Problems of Time and Space,* New York: Macmillan, 1964.

The scientific basis for the human faculties

The fundamental differences between the measure formulae, demonstrated by the different ways they are known, suggest that the origin of feeling, thought, and sensation may not be merely human, but may lie much deeper.

When we describe the human organism in this way, however, we are in double jeopardy. The behaviorist says: "You see, the human organism is merely a machine"; while the vitalist counters: "When you try to prove that man is merely a machine, you deny free will."

We would answer both points of view by referring to the fourth aspect, control. In a machine, this would include all the devices provided for its control: the accelerating, braking, and steering mechanisms. But all the mechanisms, no matter how elaborate, do not complete the control. *Control must be initiated by an operator.* And this absolutely essential element is indeterminate and unknowable to an observer.

Thus the complete account, far from denying free will, shows both the possibility of free will *and the requirements for its effective operation.* For if any of the three "mechanical" aspects does not function properly or is tampered with by an enemy agent or even a behaviorist, free will cannot manifest itself.

The learning cycle

The learning cycle further illustrates the cycle of action. An infant begins with a *spontaneous act.* He reaches out to grab something (1). Then he encounters a painful contact, perhaps a hot stove, and he *reacts* by withdrawing his hand (2). After the pain subsides, he takes stock, associates the stove with pain, *observes* the situation (3). Finally, he *controls* his action by avoiding hot stoves (4).

The four types of action in the learning cycle correspond to the four measure formulae:

Spontaneous act (impulse)	= acceleration
Reaction (also spontaneous)	= velocity (i.e., change)
Observation	= position (the observable factor)
Control	= control

The same diagram that we used to depict the cycle of action may be used to represent the learning cycle, but it is necessary to *reverse the clockwise order and to start at the left:*

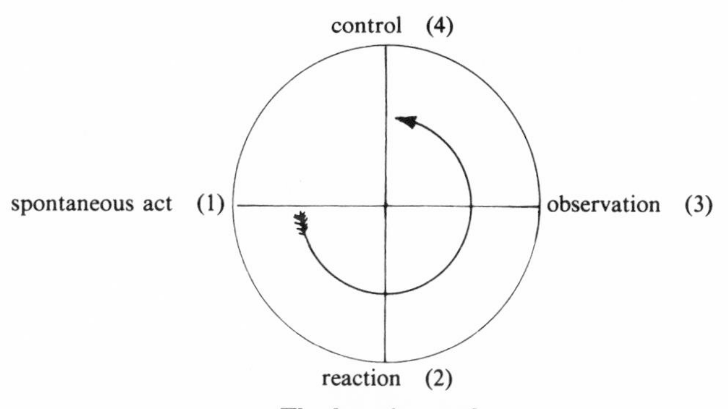

The learning cycle

While it is satisfying that the same scheme serves to represent such dissimilar activities as an operator's controlling a machine and a child's learning about the world, the fact that *the order must be reversed* is important and will be discussed later. Another indication of incompleteness is that, so far, we have dealt only with actions, each of which is part of a larger cycle which begins with a stimulus and leads to a result. The behaviorists, in their dependence on the duality of stimulus and response (a form of action), fail to recognize the necessity of a third factor: when a dog's hunger is satisfied, it may no longer respond to the stimulus of food.

But when the stimulus causes wrong action and the result is not achieved, the learning cycle becomes necessary. Thus the learning cycle occurs only when there is an obstacle in the larger, threefold cycle.

III The threefold division

In discussing the learning cycle, I implied that fourfold analysis is insufficient as a general description of process. If, for instance, the infant does *not* encounter a painful experience, but goes on happily reaching for objects, the cycle includes not only actions but other factors as well. This is the threefold cycle of stimulus, response, and result:

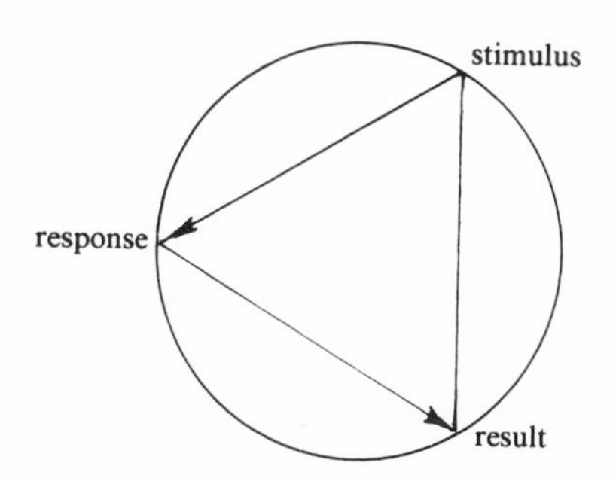

The three aspects in this case are again different categories, for which it is quite difficult to find sufficiently general names. In the case of the fourfold analysis, we've seen that response is of four kinds: action, reaction, observation, and control. We will find in due course that "stimulus" is of four kinds, and likewise "result." A stimulus is a relationship between things—

for instance, between me and my dinner—and, as we shall see later, we would do well to use the more general term "relationship" for this category. "Response" may be similarly generalized by the word "act." And "state" would be a better word than "result," because since the process is circular, what we call result also precedes the stimulus. A state of hunger is the necessary antecedent to the stimulus of food, just as a state of satisfaction will be the result of eating.

So the generalized description of the threefold cycle is:

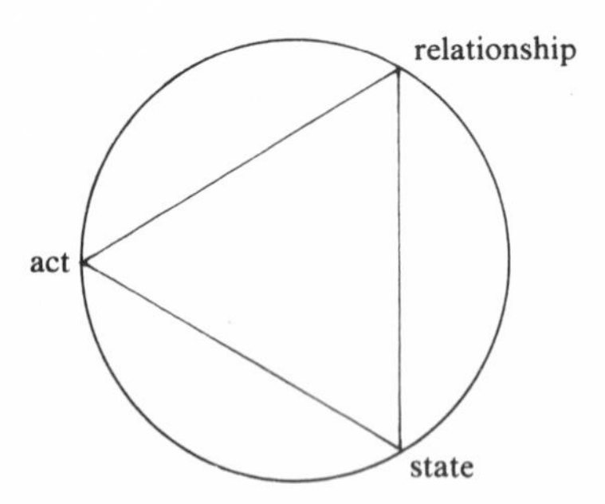

In returning to a state, which it must do to be a cycle, the process does not necessarily return to the *same* state that initiated it; it only returns to a state of some kind—in the state category. Similarly, as we said before, the position reached by the operator of a vehicle is not the position from which he started, but is in the position category.

Examples of the threefold

The threefold is not limited to this cycle. In fact, the cycle, as an analytic concept, does not fully describe the threefold. In the more general sense, this is a way by which wholes divide into three interrelated factors, often, but not always, in the form of two elements plus that which is between them. It is a way of interpreting an unlimited variety of ordinary phenomena.

Since it is basically nonconceptual, it cannot be defined, so I will try to describe it by means of a few examples.

Time

The most fundamental instance is time. Its factors are:

past ⟶ present ⟶ future

Time carries with it the idea of movement, whether we think of ourselves as moving through time, or of time as moving through us. There is a compulsive quality about time, a commitment to it, that makes it quite different from distance in space, where we can move in either direction. We cannot travel into the past as we can travel to Chicago and back.

It is the one-wayness or asymmetry of time that makes possible any ordering or temporal succession, like the two directions in which the steps of the cycle of action can occur. Without the asymmetry of time, we could not refer to "before" and "after" without specifying a point of view.

Electricity

The motion of a charge in a field is another example of the threefold. Here the three axes are important, as illustrated by the right-hand rule, which is used to remember the direction of the force *(F)* on a moving charge, or current of electricity *(J)*, in a magnetic field *(B)*. Note that field → motion → force correlates with relationship → act → state.

This example has a number of variations:

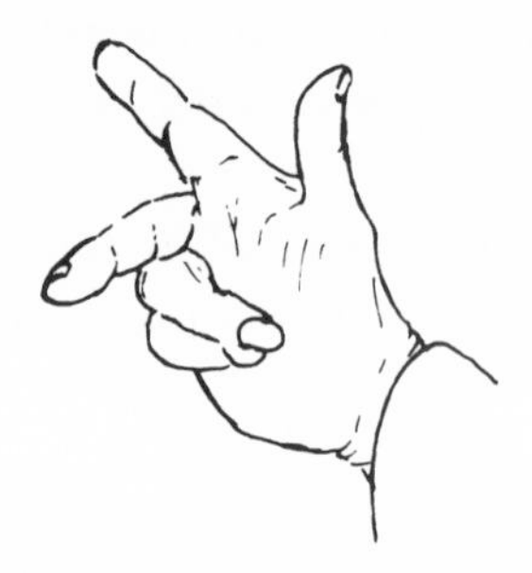

An electric motor: a moving current in a field creates a force. $J + B \rightarrow F$.

A dynamo: a force on a coil in a field creates a current. $F + B \rightarrow J$.

An ammeter: a current in a field (motion of a charge) again produces a force. $J + B \rightarrow F$.

A magnet: a current creates a field (force is required to keep the wire in place). $J + F \rightarrow B$.

These instances are *permutations* of the three ingredients—we start with A and produce C through B, or B through C, or again, we start with B and produce A through C, etc.

Stability

There are three types of stability or instability, for example, in an aircraft. These are:

1. Phugoid instability, which oscillates: Fig. 14
2. Catastrophic instability, which causes upset: Fig. 15
3. Neutral stability, which doesn't change: Fig. 16

These are illustrated by:

1. A ball in a bowl: Fig. 17
2. A ball on a convex bump: Fig. 18
3. A ball on a flat surface: Fig. 19

These three kinds of stability are due to displacement of the center of lift (CL) of an airplane with respect to its center of gravity (CG). (For the ball they are due to the relative position of the ball and the center of curvature of the surface.)

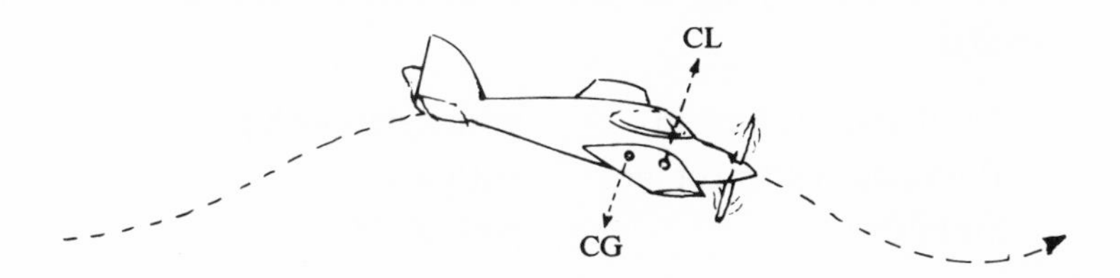

1. If the CL is ahead of the CG, measured with respect to wind direction, the airplane will dive and rise, dive and rise.

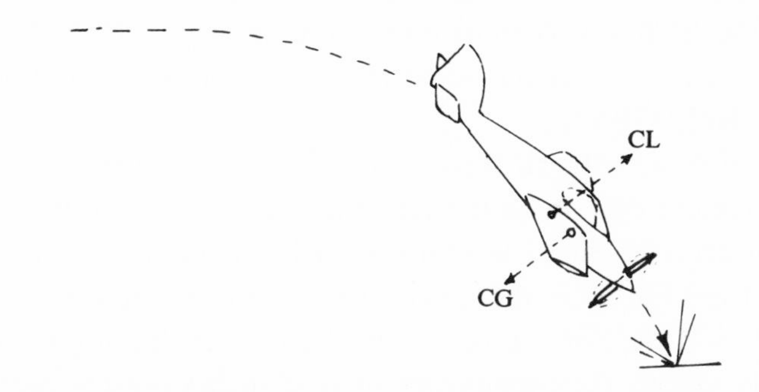

2. If the CL is behind the CG, the airplane will dive and won't right itself.

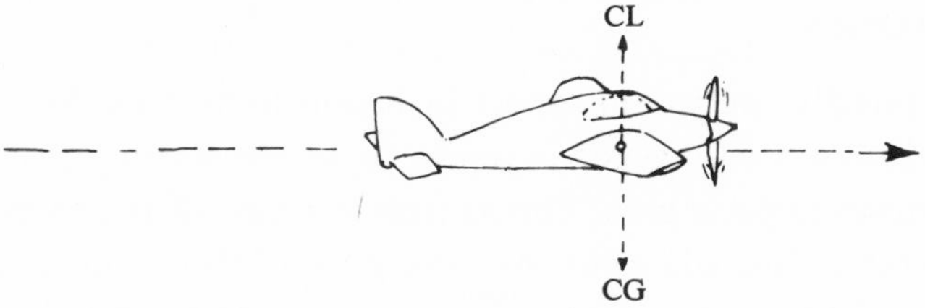

3. If the CG is on the CL, the airplane will fly level and remain so.

The permutation of center of gravity and center of lift with respect to wind direction determines the behavior. These three kinds of stability correlate to some extent with the three modes:

Oscillation (phugoid)	with relationship
Plunging (catastrophic)	with act
Stability	with state

Three-ness in cosmogony

Three-ness appears in many religious systems as a trinity of gods; for example, the Brahma, Vishnu, and Shiva in the Hindu tradition. A most interesting one is the Holy Trinity of the Christian religion—God in three persons—Father, Son, and Holy Ghost.

While, as with all threes, there is here an intrinsic movement that shifts the reference as soon as one tries to pin it down, one can at least distinguish a temporal succession. God the Father is the *antecedent* to any situation (in the past), and God the Son, *successor* (in the future), leaving God the Holy Ghost, the *immanence* of God in the present moment. It is through the Holy Ghost that God the Father conceives the Son.

Inversion

Ordinarily, we would expect inversion to be a twofold operation: two inversions return us to the starting point, two minuses make a plus. This is true in terms of the fourfold operator. But when the inversions are in three dimensions of space, it takes three inversions, not two, to get back to the start. Another interesting property is that any two inversions are equal to the third (in the illustration, 1 + 2 = 3, 3 + 2 = 1, 3 + 1 = 2).

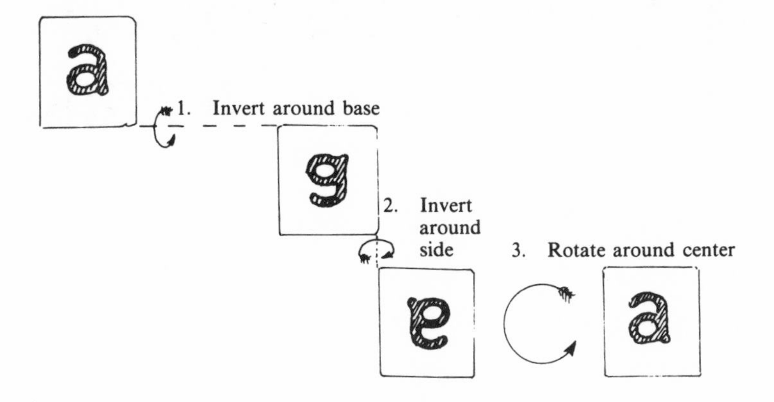

Sentence structure

Sentence structure involving subject, verb, and object could be studied as an example of the threefold. The verb, of course, would supply the activity—but only a verb in the present tense; the other tenses are not active. This is offered not as an illustration, but an area for further thought, and warns us of the subtleties of this pre-analytic stage.

IV Generating other measure formulae

In Chapter II, we noted that position and its derivatives, velocity, acceleration, and control, are measure formulae. But there are other measure formulae to which we should now give attention. In the pursuit of meaning, the measure formulae are invaluable because they replace words. Words depend upon definitions, which in turn depend upon other words, making a closed circle. The measure formulae, as we have seen in the case of length *(L)* and its derivatives, are more explicit than words in that they define operations, like the operation of the derivative. Thus they anticipate the aspects or categories that we may expect a situation to have. Automorphic operations are operations which, when applied, repeatedly return one to the start after a certain number of steps. Thus the operation of turning through an angle of 90 degrees when applied four times returns one to the start. Similarly, negation is a two operation because the negative of negative (double negative) is positive.

The measure formulae are the outcome of centuries of struggle on the part of physical science to arrive at a set of terms that can be formulated in such a way as to leave no vagueness, no ambiguity.* They describe the skeletal anatomy

*For an excellent description of these struggles, see Jammer, Max, *Concept of Mass*, Cambridge: Harvard University Press, 1961.

of all of science, and even anticipate modern discoveries. Quantum theory, in giving prominence to the quantum of *action** as its most basic ingredient, only drew attention to a concept already included in the measure formulae.

The measure formulae we have considered—position and its derivatives—are given in terms of two variables, length *(L)* and time *(T)*. These two variables, more correctly referred to as *parameters,* together with a third, mass *(M)*, are the basic ingredients of physics, more basic even than the formulae, as they are fewer in number; the formulae are combinations of these parameters. *M, L,* and *T* have the same status in physics that points and lines have in geometry; they are the "undefined terms," the fundamental ingredients from which all the measure formulae are constructed.

The formulae we have considered so far do not involve mass. If we multiply each of these formulae, length and its derivatives, by mass, we get a second set of measure formulae:**

Group I			*Group II*	
Length**	L		Moment	ML
Velocity	L/T	$\times M =$	Momentum	ML/T
Acceleration	L/T^2		Force	ML/T^2
Control	L/T^3		Mass control	ML/T^3***

By multiplication of a new parameter *M,* we have obtained a second set of four measure formulae. If we now multiply the second set by *L,* we will obtain a third set of formulae:

*Not to be confused with the cycle of action, as we shall see shortly.

**Position was our earlier term, but here we use length because it is the measure of position.

***Until recently there has been no conventional word for the third derivative of momentum. It is now recognized, however, and used in aeronautical engineering, where it is called power control. I would prefer to call it mass control (material extent of control).

Group I		*Group II*		*Group III*
L		ML (moment)		ML^2 (moment of inertia)
L/T	$\times M =$	ML/T (momentum)	$\times L =$	ML^2/T (action)*
L/T^2		ML/T^2 (force)		ML^2/T^2 (work)
L/T^3		ML/T^3 (mass control)		ML^2/T^3 (power)

These twelve formulae include all of the ten used to analyze the dynamics of a moving body, plus two (L/T^3 and ML/T^3) not presently recognized in physics textbooks, but used in engineering.

Formal device for "reducing" *M, L,* and *T*

In the physical sciences, *M, L,* and *T* are considered basic, that is, not capable of further reduction. In our search for the origin of meaning, however, we are interested in the possibility of reducing them to the even more basic unity or totality. We have already reduced a total situation (the cycle of action) into its four aspects by operation of the derivative. In the cycle of action, and in each of the other groups shown above, the operation of *T* (time) resulted in a division into four. *T,* then, may be considered as one-fourth of unity: it *quarters* the whole.

If we want to reduce *M* and *L* to aspects of the whole, we need similar conventions for representing them on the diagram. What part of totality is *M?*

I will not take the time here to conduct the reader through the process of trial and error which finally convinced me that

*The term "action" here is not to be confused with the "four kinds of action" in group I. This term, ML^2/T, is rate of change of inertia.

M is 120 degrees.* My clue was in the threefold cycle, where the shift from act to result may be seen as equivalent to *embodiment* or *incorporation* (thus "massing")—literally, as in eating a meal, and also metaphorically, as in the accumulation of experience. Action × mass = result.

Since division (or differentiation) in the case of *T* was represented by clockwise rotation, multiplication (or integration) should be counterclockwise:

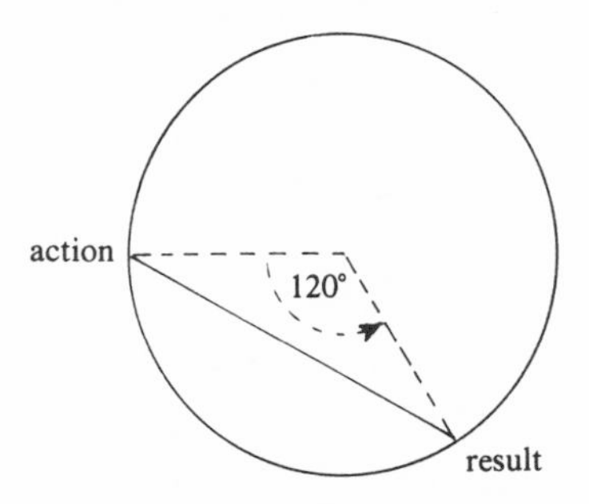

Using this device, representing multiplication by *M,* we may transpose to the diagram of the fourfold:

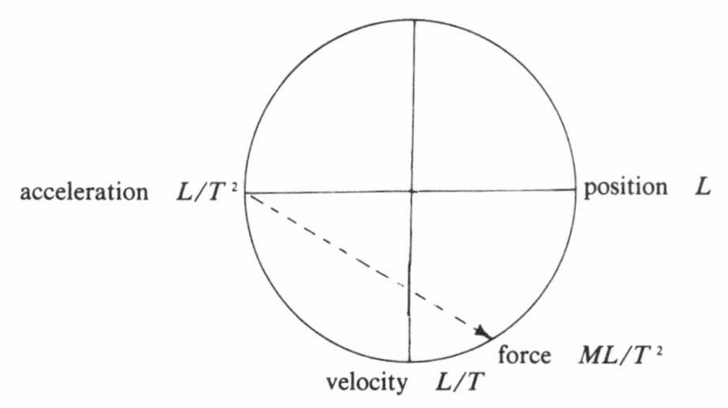

Then we may set down the remaining group II formulae on the circle representing totality.

*One hint is that mass is volume times density. Since volume is L^3, and density is a number equivalent to another *L,* we can suppose $M = L^4$. As we shall later show, $L = 30$ degrees so $M = 120$ degrees.

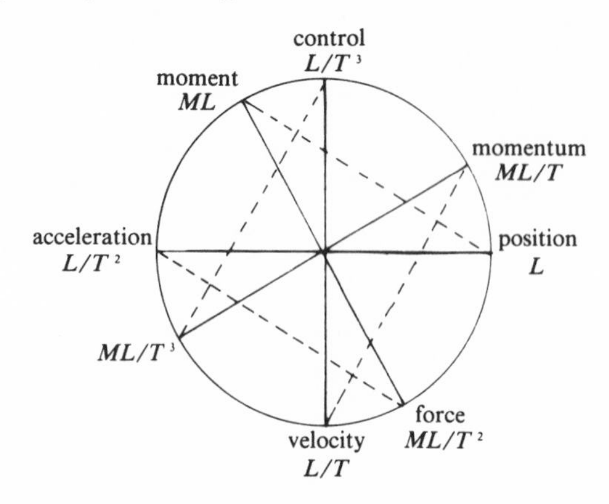

The placement signifying mutiplication by L is not so difficult, because there are now only four spaces remaining on the diagram for the four formulae in group III. Of the four available spaces, only two (30 degrees and 120 degrees) would divide the circle an integral number of times. And since 120 degrees is already used to represent M, this leaves 30 degrees, or one-twelfth of the circle, to represent L.

By putting in the formulae of group III at 30 degrees from their corresponding group II formulae, we have completed the diagram of totality in terms of M, L, and T.

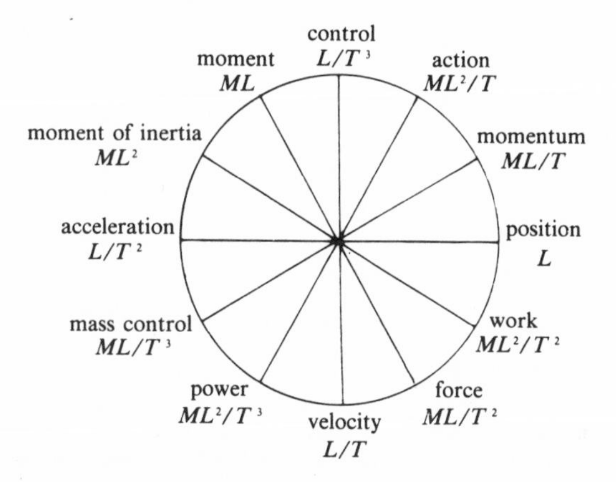

Here, then, is the division of the whole into three and four parts, referred to in the Introduction. (We will not deal with the twofold division until later.)

When I first attempted to find a division of the whole into parts, I tried to make these divisions *M, L,* and *T,* plus one more to be discovered. But this did not work, and finally it became clear to me that *M, L,* and *T* are three *different ways* by which the whole is divided: by *T* into four parts, by *M* into three, and by *L* into twelve.

These discoveries, as I have attempted to show, were accompanied by the recognition of angular relationships between the different fundamental terms, the measure formulae.

V The rosetta stone of meaning

The method developed in the foregoing chapters has shown the interrelationship underlying the complete basic vocabulary of science. Since we have already found that some of these scientific measures (in the cycle of action) have equivalents in human terms, we may likewise expect to find that the others have equivalents in human terms.

This is an area to which the specialist, philosopher, mathematician, or even psychologist, has no better access than the unskilled human being.

I should point out that the diagram we are evolving—which I call the "Rosetta Stone" because of its similarity to that famous tablet which bears inscriptions in three languages, providing the key for the translation of Egyptian hieroglyphics —is not just a translation of meaning, but is a *generation* of meaning. It is the relationships between the words we must use, not their definitions, that give them their meaning. With this in mind, we will now attempt to uncover the relationships underlying the basic human vocabulary.

The learning cycle has given us four basic categories of act:

1. Spontaneous Act
2. Change (reaction)
3. Observation
4. Control

Considering the threefold cycle, we may expect each kind of act to have a preceding stimulus and an ensuing result or, as we generalized, an appropriate *relationship* and *state.* But in Chapter I, we already disclosed four categories of relationship, so now we need only reexamine these four and see to which acts they properly correspond. We may then find states corresponding to the four acts.

Our analysis of the four relationships is best done through the two dichotomies of objective–projective and particular–general. By considering the permutations of these dichotomies, we can explore the meaning of each kind of relationship, and then look for an appropriate word to describe it.

1. *Objective general.* General information which is objective, definitions, scientific laws, etc., we will simply call KNOWLEDGE, in the sense of a "body of knowledge."

2. *Objective particular.* Particular objective information, on the other hand, such as "This triangle is dented," is factual. We will call this category FACT.

3. *Projective general.* Knowledge which is projective and general, such as "All redheads are talkative," might be called belief. Since "belief" suggests a particular situation, however, we would use the more generalized word FAITH, remembering that the projection can be correct or incorrect.

4. *Projective particular.* This category is inherently subtle and difficult, containing relationships within the person himself. For the moment, we will leave it unnamed, but later we will find that it is possible to name it.

We must now decide which of the four acts corresponds to each of these relationship categories, and what would be the state resulting from each act.

1. KNOWLEDGE. The act immediately appropriate to knowledge (as it is meant here—a body of knowledge) is OBSERVATION. By observation we mean not only "looking," but other kinds of consideration as well. A body of general knowledge is useless until it is considered.

   The state immediately resulting from the consideration of knowledge is SIGNIFICANCE.

2. FACT. The act appropriate to a particular fact is CONTROL. Upon encounter with fact—the traffic light is red—one controls the situation by stopping the car. Recall that control cannot become manifest without certain objective facts at its disposal.

   The word I have tentatively chosen for the state resulting from control is ESTABLISHMENT. It could also be called "accomplishment" or "consolidation."

3. FAITH. Because faith is a presumption or an expectation, it is likely to produce CHANGE (reaction) upon confrontation with actuality: I meet a redhead who isn't talkative. (Change here is in the passive sense of *being changed;* the active sense, *producing change,* would be "control.")

   The state immediately resulting from change is the TRANSFORMATION of one's original faith.

4. IMPULSE. This is the name which, in deference to our method, we refrained from assigning to the "projective particular" relationship. Now we can see the difficulty: what is the antecedent to a "spontaneous act"? Here we must think of *spontaneity* as being in the relationship category, where we will call it "impulse." Other possibilities would be "insight" or "intuition." Since a spontaneous act is projective, directed toward the future, its stimulus is not apparent. It is equivalent to purpose (within the person himself).

   The state which results from the spontaneous act, for instance, a playful act, is very simply BEING.

The foregoing is summarized thus:

Relationship	*Act*	*State*
Impulse (purpose)	Spontaneous act	Being
Faith	Change (reaction)	Transformation
Knowledge (form)	Observation	Significance
Fact	Control	Establishment

To complete the "Rosetta Stone," we must show the correspondence of these twelve human categories to the twelve measure formulae. We shall use the cross axis as a format for analysis to bring out the meaning implicit in the angular relationship between the terms. Recall that factors at opposite ends of an axis are mutually opposed, that those at right angles to one another are independent, and that rotation signifies change in time.

The four relationships

Impulse/Action (ML^2/T): In physics, the quantum of action exists in its pure state as a quantum (discrete particle) of radiation, having an undividedness which makes it unique among physical measures. These "quanta of action" are the origin of matter, and in having the power to alter the state of an atom, they are the origin of change.

Impulse, like the quantum, is the *initiating factor* in a process. Also like the quantum, it is instantaneous and quantized—occurring in discrete units. Two of its forms, decision and recognition, emphasize this—you cannot recognize somebody, nor make a decision, one and a half times.

Impulse and action are particular, potential, unknown, projecting into the future.

Faith/Moment of inertia (ML^2): In science, the moment of inertia is the tendency of a thing to continue in a given state of rest or motion. In mechanical systems, it serves the function of maintaining steady motion and canceling out fluctuations, as with the flywheel of an engine.

Its human equivalent, faith, is the tendency to maintain a given credo without examination. Like the flywheel, it serves to maintain steady motion, carrying us through the vicissitudes of life.

Faith and inertia are in the present. Faith is the projection we put upon the present situation.

Fact/Work (ML^2/T^2): Work is energy—the amount of energy expended to perform a task. It is opposite to inertia.

For the mind, work is the readjustment of its implicit beliefs when confronted with fact. Like work, it is a physical exchange of energy, as when a scientist tests his theory on physical objects.

Work and fact are in the present—they are the impingements of the physical world upon the person. Fact is equivalent to work because the word is derived from *factus,* past participle of *facere,* "to make," as in "factory."

Knowledge/Power (ML^2/T^3): Knowledge, or data, is objective descriptive information which sits in the library or in our minds until it is used.

Power is an objective measure—the description of the dimensions of an engine, for example. By itself it is nothing; power has no actuality until it operates for a time to produce work. Power $\times$ time $=$ work.

Knowledge and power manifest through application.

The four relationships

The foregoing diagram should be compared with the four types of relationships outlined earlier. It is worthwhile to play around with the interrelationships involved in order to get a sense of how together they constitute the totality of the object.

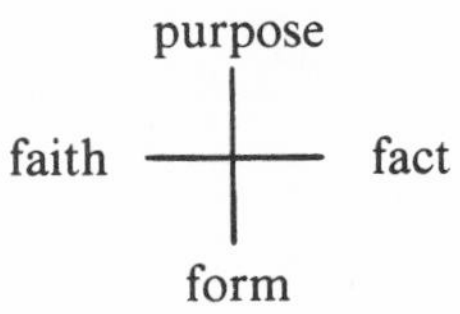

purpose

faith — fact

form

Note that faith and fact are opposite. This does not imply that all belief is false, but that the element that operates in faith is a projection of what has gone before. Like inertia, faith is the continuation of the state of rest or motion that has held before, whereas fact is the confrontation with what is particular to the present.

Similarly, function and form are opposite. Function is the purpose which the self projects for the object (I use a stone for a hammer). Form is the definition that would limit this freedom (I ask for bread and you give me a stone). Again, we may note that either form or function may mediate to determine whether something is or is not what it is called. Are these your glasses? No, mine are bifocal (form or definition decides). No, but they will do (function decides).

A difference between vertical and horizontal is that the horizontal is bound to time. It concerns what is immediate or phenomenal. The vertical axis is not bound to time. It concerns the ideal, either the definition, which is ideal in the sense of a standard for manufacture; or the function, which is ideal in the sense of the cause that gives the object its value.

The four acts

Control/Control (L/T^3): Control is the final stage of the learning cycle, and in a moving body it is the rate of change of acceleration. Control is free; it is at the disposition of an operator and correlates with *will.*

As the reaction to observation, it is *conscious action.*

Spontaneous act/Acceleration (L/T^2): The spontaneous act is primary and simple, without antecedent. In motion, the initiating factor is acceleration, as in starting a car from rest.

Both the spontaneous act and acceleration are *unconscious action.*

Observation/Position (L): The midpoint in the learning cycle is observation—the consideration of what has taken place. For the scientific term, we may equate it with the thing observed: position.

All the other measures ultimately involve a measure of observable position, such as a needle on a dial.

The observation of position is a *conscious reaction.*

Change/Velocity (L/T): Velocity is the change of position, and we may generalize it as the change of any observable property. In the learning cycle, it is the point of reaction to the initial act.

Velocity and change (passive) are *unconscious reaction.*

The four acts

Here too there is value in drawing out the implications of opposite and of complementary aspects. It is clear that observation on the right is opposite spontaneous act on the left, and it is also clear that control at the top is the opposite of the passive change at the bottom of the chart.

This oppositeness is confirmed by the literal oppositeness, in the case of the pendulum, of acceleration and position. When the pendulum is at the extreme position, acceleration is at a maximum and is pulling in the opposite direction.

If one experiments with a pendulum hung from the hand, it will be found that one can control the pendulum by timing the motion which produces control so that it is exactly opposite to the velocity. In other words, control is most effective at the midpoint of the swing. Here a control motion in the direction

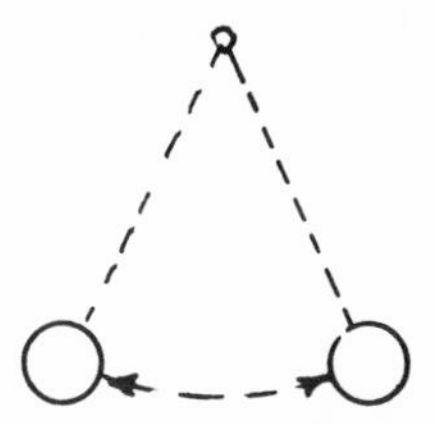

of the pendulum will stop it, and a motion in the opposite direction will increase its swing.

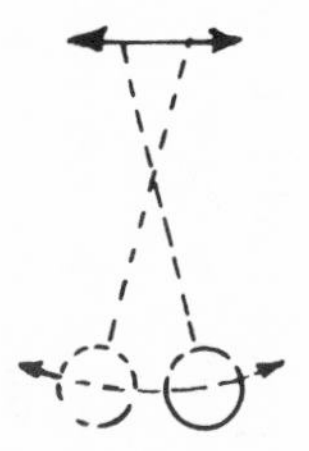

Such control motions, which do not have to lift the pendulum, can be effected with a minimal energy reduceable in

The four states

Because they are not conceptual, the human states must be apprehended through feelings. While each physical state is the rate of change of the one before, it would appear that the human states do not change into one another except through the appropriate relationships *and* acts; *and we are reminded that the fourfold is essentially secondary to the threefold.*

Significance/Moment (ML): These two terms come together when we refer to "matters of great *moment,*" meaning significance. In a broader sense, it is the extent to which things *come into focus* on an issue. In science, moment is leverage—the "state" of a mass at a distance—as when a man uses a crowbar to move a stone many times his own weight.

Establishment/Mass control (ML/T^3): Recent aeronautical practice refers to this product of mass times control as "power control," but strictly it is "force control" or "mass control." One can recognize the difference between controlling a small mass, say, a pencil, and bringing a ship into dock. The control of a greater mass is a greater accomplishment. "Establishment" is a tentative general name for this product. We could also use "accomplishment" or "consolidation."

Transformation/Momentum (ML/T): By "transformation," we mean the state resulting from a change—a certain condition within an overall "state of flux." It is distinct from "establishment," which is a final state of control, usually changeless. Thus the woman admiring her new hairdo is celebrating a transformation, rather than the possession of beauty. In science, this "state of changing" is called "momentum," as the momentum of the hammer drives the nail, or the momentum of the car breaks the telephone pole.

Being/Force (ML/T^2): Like "impulse" and "spontaneous act," "being" is elusive as a concept. Insofar as it can be objectively described, it is the result (within the actor) of a spontaneous act—the *incorporation* of spontaneous action. It is helpful to contrast it with "having." It may be more easily understood in its scientific guise as *force,* a condition which is caused simply by the presence of something—as the force of gravity is caused by the presence of the earth. So "being" is like the *force* of personality.

theory without limit. This, of course, confirms the independence of the control aspect and the motion aspect, because were the control not independent of the motion, it would not be possible to induce motion in either direction.

The four states

The four states can be seen most simply as the dichotomies of being and nonbeing and of having and not having.

Establishment, or mass control, is easily correlated to having, but many things that one may have are not desirable, for example, sickness, impediments, flaws. Transformation is the state in which such undesirable limitation has been

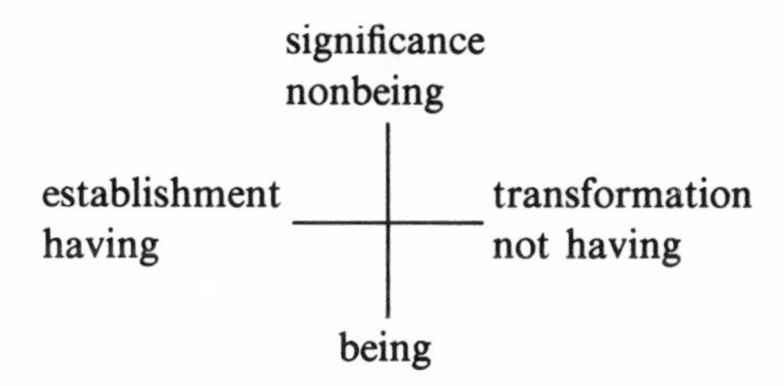

removed (not having). Transformation includes the destruction necessary to construction; in addition, it includes the breakdown of waste products, making them available for new growth.

Nonbeing is consciousness of being because it is opposite to being. But to be conscious of being is to realize the significance of being; hence, nonbeing is significance.

This use of negation (as in giving importance to nonbeing) is perhaps awkward in comparison with the usual words, but it is a technique that can handle meaning and can reduce two words to one or even four words to one, if we could find word pairs as we did for relationships and action.

The four states are difficult because they are, as noted, not conceptual, nor are they apparent in the sense that actions are. It is especially important to recognize that states are represented in the measure formulae, and hence have status in science coequal with the relationships and actions. So great is their importance in current science, in fact, that it is sometimes stated that science consists solely in the observation of states.

We have had examples of actions and of relationships, so we should have one of states. We can think of the four stages of an internal combustion engine as bringing about four different conditions, or *states,* of the contained gases.

First there is the compression stroke in which the mixture of fuel and air is compressed, second the firing stroke in which the gases explode and push the cylinder to do work, third the exhaust stroke in which the waste products are discharged, and fourth the intake stroke in which new fuel and air is drawn in. These follow the sequence in counterclockwise order:

Compression	Gases compressed	(Establishment)
Ignition	Explosion and expansion	(Being)
Exhaust	Wastes discharged	(Transformation)
Intake	Gas is drawn in	(Non-being)

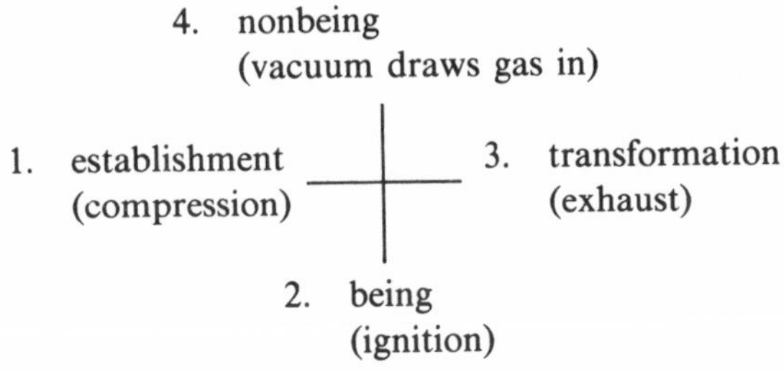

4. nonbeing
(vacuum draws gas in)

1. establishment
(compression)

3. transformation
(exhaust)

2. being
(ignition)

The rosetta stone

Physical quantities correlated with their equivalent English meanings

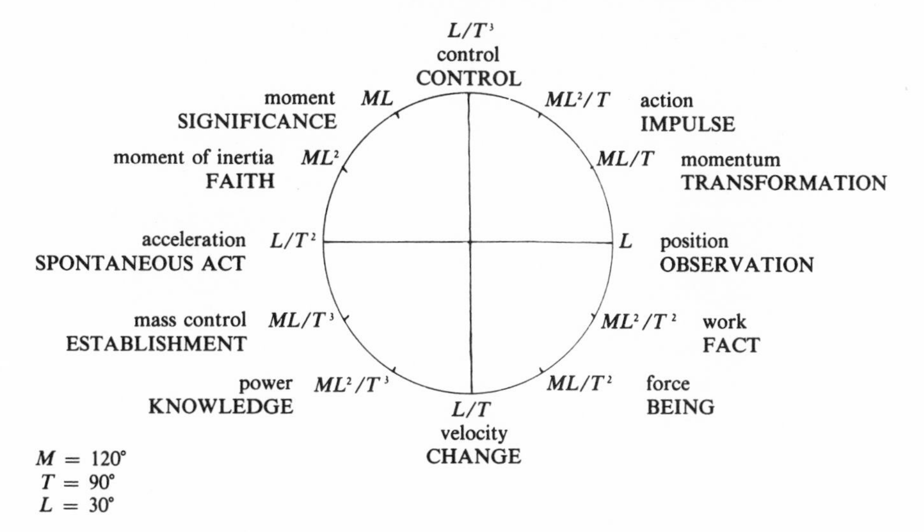

The diagram above is the same as that at the end of Chapter IV, except that to it have been added the words that we have been dealing with in this chapter. As with the measure formulae, we can move around the circle in threefold or fourfold manner.

We have now completed the diagram by adding the four stimuli and the four results (or relations and states).

We should also realize—a point which will be discussed in more detail later—that there are two directions in which to go around, counterclockwise and clockwise. The motion is counterclockwise for the learning cycle, moving from blind action (acceleration) through reaction to control. It is clockwise in the operation of a machine because the operator anticipates what happens.

The learning cycle is natural or naïve, but teaches control; the other is informed and applies the knowledge gained.

Illustration from Peirce

I will close this chapter with an analysis by the philosopher Charles Sanders Peirce (referred to in the Introduction). In his essay "How To Make Our Ideas Clear,"* Peirce describes four steps which lead to action. We begin with *sensations,* of which we are immediately conscious. These, he maintains, occur in succession and create a *thought,* just as the succession of musical notes creates a melody. The goal of thought is *belief;* we continue the activity of thought until we reach a belief, the "demi-cadence which closes a musical phrase in the symphony of our intellectual life." He goes on to say that belief establishes a rule for *action,* so that the final upshot of thinking is the exercise of volition.

Thus we have the sequence: sensation → thought → belief → action. This is precisely the sequence we would have on a diagram of the four relationship categories if we were to proceed clockwise from "fact":

Here, as with the aspects of motion, each right-angle shift is the slope or trend in the previous category. With relationship, we have, first, awareness of isolated facts; then, with repetition of fact, knowledge of generalities; then, confidence in such knowledge; and finally, we decide on action based on this confidence.

This succession moves in the direction of a conscious search for valid principles of action (clockwise); it is opposite to the learning cycle (counterclockwise), which advances by trial and error.

Piaget has found this sequence in children. Over a period of years, they become able to (1) move from observation to recognition of law, (2) act on the assumption that law holds despite counterfactual evidence.

**Philosophical Writings of Peirce* (p. 314). Justus Buchler, ed. New York: Dover Publications, Inc., 1955.

VI The roots of unity

Negative and imaginary numbers

I will now venture into another discipline, algebra. The use of negative and imaginary numbers in algebra confirms the fourfold nature of analysis, and also provides additional insight into the threefold. Here again, I should acknowledge that I am using the concepts of a discipline for purposes outside their usual application, but there are certain discoveries of mathematics which make valuable contributions to our study.

The evolution of mathematics was given great impetus by the discovery that it was possible to use negative numbers, *negative quantities.* If we represent positive numbers extending to the right of zero, we can represent negative numbers extending to the left:

$$\text{etc.} \ldots -3 \; -2 \; -1 \;\; 0 \; +1 \; +2 \; +3 \ldots \text{etc.}$$

With this device, we may describe addition as moving to the right, and subtraction as moving to the left. This makes it possible to subtract a larger number from a smaller one; for instance, if we take 3 from 1, we get -2, which is a real (although negative) quantity.

Another important concept was that of imaginary numbers. They were not so much discovered as encountered.

Mathematics had arrived at the concept of a number as having roots; numbers which, multiplied together, will produce that number. When the concept of negative numbers came along, there was a clash. What would be the two numbers which multiplied together would produce a negative quantity, -1, for example? For a time there was no answer. The square root of a negative quantity must be impossible. So it was called imaginary. But when Gauss, called by Bell the prince of mathematicians, found a method for representing imaginary numbers, it was not long before their value was appreciated, and today they are just as important as real numbers. This method uses the Argand diagram, which, in essence, correlates unity to the circle, and roots of unity to fractions of the circle.

Recall that negative numbers were pictured as extending in a direction opposite to positive numbers. In this way, the square roots of unity, $+1$ and -1, can be expressed as the opposite ends of a line with center zero. This line can be thought of as an angle of 180 degrees, or a diameter.

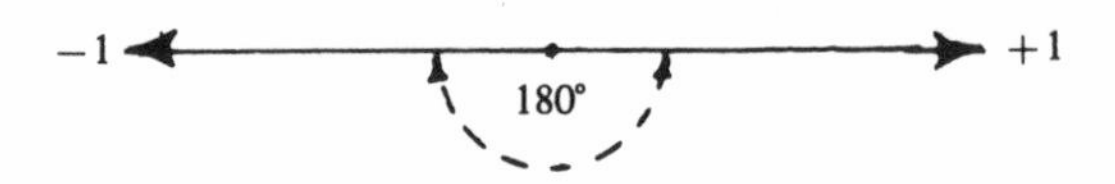

Gauss extended the idea further and pictured $\sqrt{-1}$ as halfway between $+1$ and -1, or an angle of 90 degrees from the line -1 and $+1$. Thus, if the division of unity into plus and minus is a diameter, or 180 degrees, a second division leads to an axis which "mediates" this diameter, or an angle of 90 degrees.

Thus we have two axes—the horizontal representing positive and negative real numbers, and the vertical representing positive and negative imaginary numbers. These two axes form the complex coordinate system, and a number on the plane

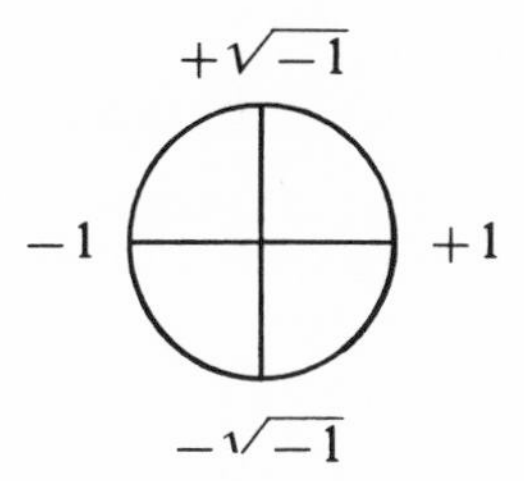

described by these axes is a number having a real part and an imaginary part.

Using the Argand diagram, this circle of unit radius (radius = 1) on the complex coordinate system, the other roots of unity (cube roots, fifth roots, etc.) are found simply by dividing the circle into three, five, etc., equal parts. Finding the roots of unity becomes simply a matter of inscribing polygons within the unit circle: a triangle for cube roots, a pentagon for fifth roots, etc. The roots are the points on the circle; their values have a real part and an imaginary part, and are measured along the horizontal and vertical coordinates respectively. This means that they are measured *in terms of square roots and fourth roots.*

From this extremely powerful simplification, it follows that all analysis is fourfold—any situation can be analyzed in terms of four factors or aspects. This not only confirms Aristotle (his four causes) but explains why quadratic (literally, "four-sided") equations occur so frequently in mathematics.

But the important generalization that all analysis is fourfold works both ways. It shows both the extent of the fourfold and the limitations of analysis, for there are things in the content of experience that are beyond analysis.

Staying within the geometrical method already set up, we can show that these nonanalytic factors involve three-ness, five-ness, and seven-ness. Despite the fact they can be described analytically, this description fails to capture their true nature.

Cube roots and the threefold operator

As I said, we can express the cube roots of unity *analytically* in terms of *square* roots; that is, using the two-dimensional diagram.

These roots are obtained by inscribing an equilateral triangle within the circle. One of these roots is $+1$; the other two are points half a unit to the left of the vertical diameter, and half a *side* (of the triangle) above and below the horizontal diameter. Since the side of the triangle is $\sqrt{3}$, the vertical coordinates are $+(i/2)\sqrt{3}$ and $-(i/2)\sqrt{3}$, the i being used to indicate that we are measuring in the vertical direction. (The imaginary $\sqrt{-1}$ is commonly represented by i.)

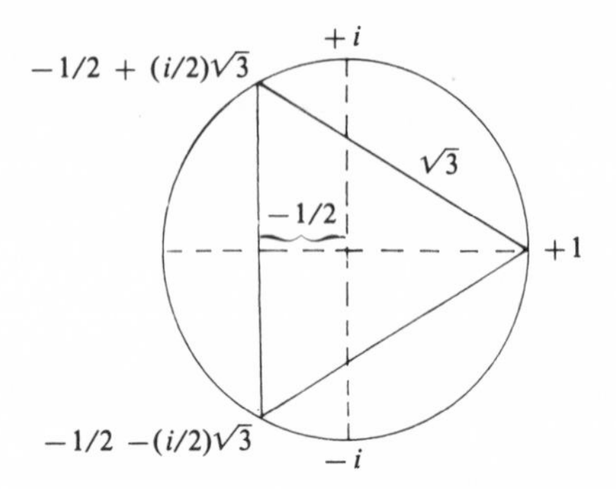

By this we see that the values for the *cube* roots of unity can be expressed as *square* roots. But $\sqrt{3}$ is an irrational number, meaning that it is not a ratio of whole numbers. Since $\sqrt{2}$ is the diagonal of a unit square,* we might expect to find some expression for $\sqrt{3}$ in units. We need not look far to find what this is, for $\sqrt{3}$ is the diagonal of the *unit cube.***

So in order to represent $\sqrt{3}$ in unitary fashion, *we must leave the two-dimensional plane.* Full representation of the

*A unit square is a square whose side is 1.

**A unit cube is a cube whose side is 1.

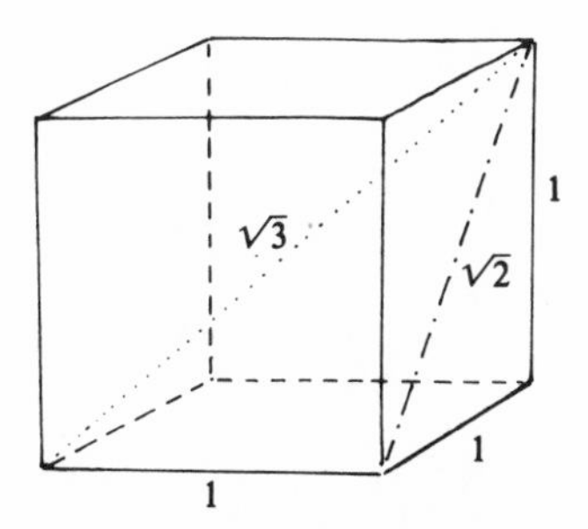

cube roots of unity ultimately involves the three dimensions of space. The threefold operator, *represented analytically* as equidistant points on the circle, is actually a three-dimensional *activity,* whose measure gives only its analytic aspect. The analytic aspect, which is in two dimensions, does not convey the full meaning of the cube root; it is like the shadow of a solid figure.

The threefold nature of the cube root is nonanalytic. It involves categories which differ from one another more profoundly than those of the fourfold.

Here we have a formal device to show the inadequacy of analysis for a complete account of the world. This is but one example of the fundamental distinction between the threefold and fourfold operators, a distinction so important to our theory and to life that I will devote the next chapter to a comparison between these two operators.

VII Comparison of threefold and fourfold operators

An important property of the threefold operator is its nonfiniteness. This property comes about because *nonfiniteness is simpler than finiteness.*

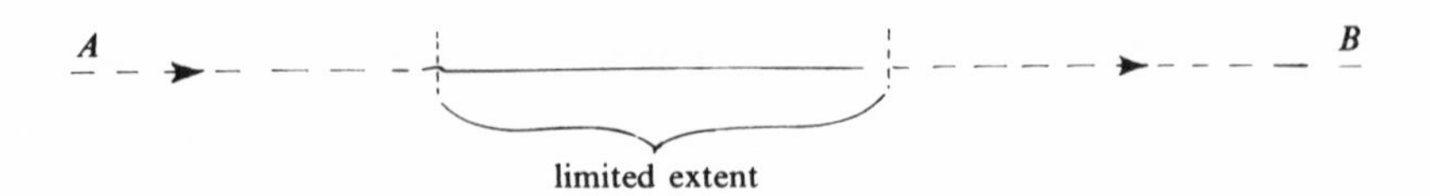

We may think of the threefold as an endless line. Between any two points on this line, we can say which is before, but we cannot look at a piece of it without getting off the line.

Any cut-off line limiting the infinite extension *A–B* (in fact, any indication even of position on the line) requires another dimension crossing the line. This new dimension, plus the line itself, takes us into a different world, the two-dimensional space of measure and analysis (the fourfold).

The threefold is but *one* dimension, flowing perpetually in one direction like a point on a line, its past having one kind of existence, its future another, and the point itself a third.* This

*In view of the one-dimensionality of the threefold, it may seem odd that we found three dimensions necessary for its representation. This can be reconciled, however, by

is experienced time, prior to clocks and measure. It is a reality which is always with us, yet perpetually eludes analytic description. Why, we might ask, does scientific description give it so little attention?* It is the old story: since it has no contrary, it cannot be isolated and subjected to analysis. Though we pour it from vessel to vessel, we never see it move.

If we try to analyze what it is that the threefold describes, we are in a bind, for it is just that element of participation in life that analysis cannot, and does not even pretend to, cope with.

If the threefold describes the "active element," it does so only by pointing to it; it does not describe it in the way a map describes the relation of points on a surface. The description given by a map is not a substitute for the relationships of points described; it *is* that relationship. The relation of points on a map is less ambiguous than the relation of places in the landscape. On the other hand, the definition of love as "the attraction of one person for another" also describes a relation,

---

thinking of the line as curving through space in an arbitrary fashion. It remains a one-dimensional line, even though its domain has three dimensions. As we will show later, this line is a constraint in one dimension but is free in two, whereas the plane is a constraint of two dimensions but free in one.

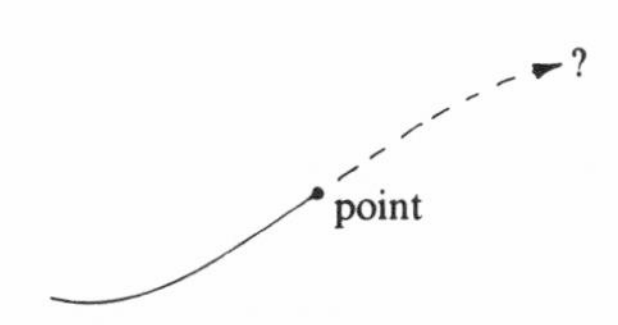

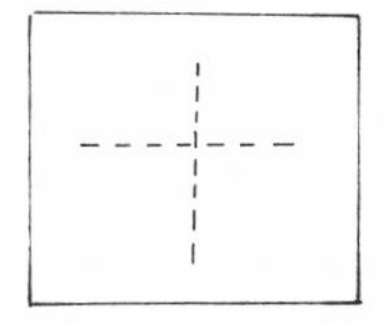

*As a matter of fact, science does recognize the asymmetry of time in the second law of thermodynamics, which states that entropy is positive (energy tends to become more evenly distributed). This is a sort of "black sheep" in the family of scientific law, both because it makes time asymmetric and because it cannot be deduced theoretically.

but the definition is completely devoid of the content of actual love. In the case of the map, the lack of content is irrelevant.

The fourfold expresses the aspects of a situation. But the shift from one mode of being to another, from action to the result toward which the action is directed, is the realm of the threefold. What it describes or refers to is the interconnectedness of modes of being for which categories or names of any sort are only keys, and do not provide sufficient description.

This is why the Zen master, when asked the nature of the Buddha, strikes the student on the head with a stick. No words can convey participation or the realization to which participation can lead. The words, the analysis, are a screen through which action penetrates (see the second diagram below) bringing the action into a new state. The threefold "stands for" all this—not in the sense that it defines it, but in the sense that its intrinsic dynamic is of the same order, time-like rather than space-like.

Two models come to mind for illustration of the threefold:

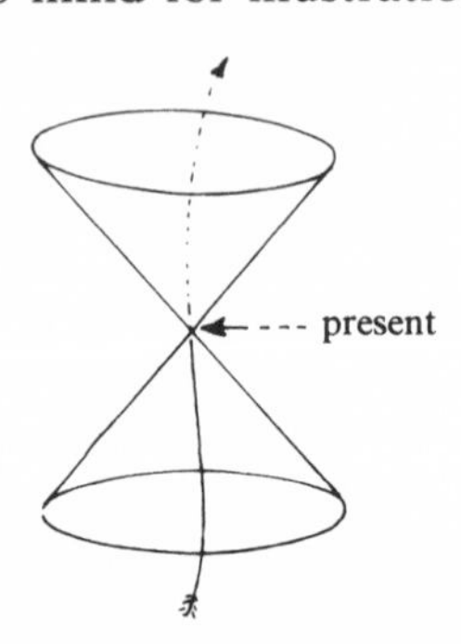

This hourglass model has the merit of emphasizing the uniqueness of the present. But the plane of qualification penetrated by the line of motion is an equally valid representation, and has the merit of showing the independence of the time line from the two dimensions which describe the plane.

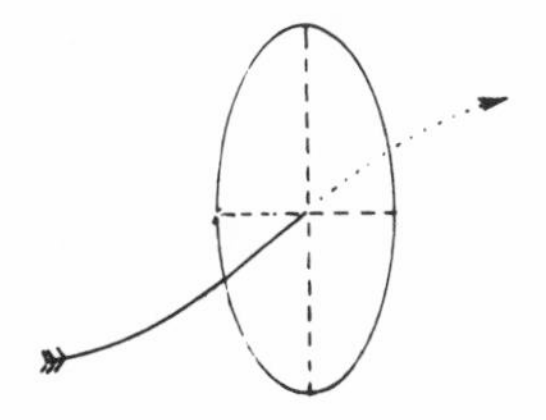

Example from psychology

The study of perception provides a good example of the distinction between the three- and fourfold. If I draw a diagram on a sheet of paper to show what a square is, what actually happens is a series of occurrences, point events in time. The *square* itself is a different kind of thing, and exists only after we have integrated a series of instances and fitted them into a meaningful pattern. The square has no real existence, any more than the Big Dipper has real existence. The eye does not see the square; rather it follows the line that describes it, as did the pen that drew it. Something other than the eye—call it the mind—integrates this motion, compares it with other known shapes, and decides it's a square.

Von Senden* collected reports on a number of persons who had been blind since birth and then received their sight by means of an operation. He found an inability to detect the difference between a circle and a triangle. Months of training were required before the difference could be recognized. This indicates that the image transmitted by the eye must be integrated, and that this process must be learned. We do not even need to say by what. We can *define* mind as "that which integrates sensation and form."

Direct experience, the threefold, is prior to perception.

*Von Senden, M. *Raum- und Gestaltauffassung bei operieten Blindgeborenen vor und nach der Operation.* Leipzig: Barth, 1932.

This supports Kant's description of time and space as *a priori.* However, there is no basis for assuming that it is time or space *separately* that is *a priori;* it is rather the linear succession of events in time that is *a priori,* or, as I would prefer to say, accompanies experience. This may be called subjective time or it may be called pure time.* But when the eye follows the outline of a square, the experience is one of movement, of time and space undifferentiated. Even if the person feels the outline of a square or walks around one, the immediate experience is a series of events in subjective time. The only order he can draw upon (when he comes to integrate) is this sequential time ordering, which does not become measurable or "space-like" until the *mind* compares one experience with another.**

I prefer to call this time flow *extension,* meaning that it is the *a priori* basis for measure of *either time or space.* As an aid to abandoning the notion that time and space are independent realities (whereas, in fact, they are only separate *concepts*), recall that almost all primitive measures of distance are in terms of time. A place is so many days' journey, etc. One measure of distance that I particularly like is used by the Yakuts, who inhabit the tundras of northern Siberia. Their unit is "the distance a man can walk out and back while a pot is coming to a boil."

Constraint and freedom

There is another important distinction between the three- and fourfold which is implied by their one- and two-dimensional natures.

*Time as directly apprehended before measure or analysis.

**Such comparison requires that we look at two things simultaneously. The matrix of this simultaneous difference is *space.*

Suppose I say, "I'll meet you at 18th and Walnut." I have specified a place but not a time. Two dimensions are committed because "18th and Walnut" is a position on the earth's surface (latitude and longitude). Commitment in space imposes two degrees of constraint.

Only if I specify both time and place can our meeting occur in fact. Three degrees of constraint are necessary. Three dimensions must be committed, one in time and two in space.

In view of the three-dimensionality of space, one might suppose that it requires three dimensions to describe position. It can be shown that one of these three dimensions is unnecessary, but I will not take the time here for the proof.*

The notion of degrees of constraint may seem to be opposed to the more commonly used degrees of *freedom.* We say the two dimensions needed to describe form are constraints, while it is usual to think of the plane as having two degrees of freedom. Actually, there is no conflict here. The point is that the plane has two *dimensions,* which may be either free or constrained; when a position is specified, two degrees are constrained. Similarly, the line has one dimension which, when specified, imposes one degree of constraint.

The threefold and fourfold operators, then, may be thought of *as using up* or *removing* the freedom implicit in the three dimensions. The combination of three- and fourfold applies to the physical space–time world of objects and events. This is the world of determinism, subject to all the laws of classical physics.

*To intercept an enemy missile, for example, we would use two dimensions to point the gun (a horizontal and a vertical angle). The third dimension in a radial direction combines distance with time of flight and doesn't need to be separated into time and space.

We may diagram the situation thus:

Level	*Character*	*Freedom*
I	Unity	Three dimensions free
II	Threefold	One dimension committed, two free
III	Fourfold	Two dimensions committed, one free
IV	Twelvefold	Three dimensions committed, determinism

The threefold has one dimension committed (time). The fourfold has two dimensions committed (mental space). The twelvefold has three dimensions committed (actual space–time).

Order of the realms

Part of our quest has been to find an order. We found that the four aspects were equally important to a total situation and hence did not establish an *order.* We then took up the basic parameters of physics, *M, L,* and *T,* whose combination produced the measure formulae. Using these fundamental "simples" of physics to divide the whole in different ways provided the order we require. And the implications are plain.

Threefold precedes fourfold, and both precede their conjunction in the twelvefold realm of physical matter. Applied to the ontology of human existence, experience precedes mind, and both precede sense data.*

*This statement seems both to confirm and deny Hume, who said that sense experience preceded mind. Our position is that experience is essentially internal and *a priori,* i.e., pleasure and pain are induced, not produced by external events. And it is only after mind has constructed itself (by integrating experience, i.e., association of pain or pleasure with specific objects) that sensation, in the sense of information from the outer world, becomes possible.

Summary

To clarify the distinction between the threefold and fourfold operators, as applied to ontology, I have listed them in a comparison chart:*

Threefold (experience, feeling)	*Fourfold (concepts, intellect)*
Indefinite, infinite	Finite
One dimension committed, two dimensions free	Two dimensions committed, one dimension free
Asymmetrical	Symmetrical
Irreversible	Reversible
Exists in time (sequence)	Exists in space (simultaneous)
Ordering can be taken only one way	Ordering possible in any direction
The matrix of experience	The matrix of mind
Supplies substance,* motion, and value	Supplies measure, form, and concept
Requires three dimensions for full representation	Requires only two dimensions for full representation

We may sum this up by saying "Time supplies content—space supplies measure." The magnitude of space depends ultimately on time to traverse it. The *measurement* of time depends on treating it like space.

A philosophical overview

Despite glimpses we may have had of its applicability or use, the device we have been constructing still seems rather aloof,

*Substance will be described in Chapter VIII.

incomprehensible, extraneous, remote from the predicament of men, whose experiences bear little resemblance to geometrical exercises. What use can there be in these abstract angles, these imaginary sections of the real that turn what is vivid and alive into something like a Renaissance drawing of figures and buildings in perspective, subservient to mere geometry?

One reads in science fiction of a visitor from outer space landing on this planet and constructing mysterious machines with rays of light, or, in old tales, of magicians drawing magic circles, waving wands, or conjuring Beelzebub with figures drawn in sand.

Why is it that virtually all of the highly developed cultures of the world have used as their symbol some geometric construction—be it the mandala of ancient Hinduism or the interlacing triangles of the Star of David?

Somewhere within us there is an instinct for this abstract world—a world that is in some sense outside life, yet upon which all life pivots; that supports it like the pins of a great steel door to a bank vault, so that if one but had the key, all would be open to him. This abstract world is at the core of our existence.

So let us have patience with the super-mathematician in us, so remote, so aloof, whom no noise or bluster can awaken, but who, when as Eddington says, "we announce we have found a group . . . of operations . . . some of which are square roots of minus one . . . ,"* begins to sit up and take notice.

How may we show that there is any essential authority in the super-mathematician? Since our thinking has been based upon the meaning of the right angle, I might try to answer this question by citing an observation which struck me when I looked up the word "right" in the dictionary. The first two definitions of "right" are: (1) straight, not crooked; (2) upright,

*Eddington, Arthur S. *New Pathways in Science.* New York: Macmillan, 1935.

erect from a base . . . perpendicular.

What struck me about these definitions is the confusion between the moral sense of the word and the geometric sense. In fact, "straight, not crooked" could apply equally well to a line on a paper and a person's moral character. I take this confusion as a manifestation of a similarity in meaning which lies deeper than the word itself. This can be seen in the word "upright," which has connotations that are not exclusively geometrical. (Up and down are, from the point of view of geometry, relative.) In order to stand up, man has to exercise a continual counterpoise of muscles. His judgment is necessary to determine deviation from a vertical position, and he must exercise this counterpoise in order to remain "upright."

So then we can at least accept that the uniqueness of the right angle, in creating a dimension that is independent of an activity under consideration, provides the measure, the *mean,* and we can now say (in the light of four-ness) the *meaning* of the activity.

Our geometry of meaning may therefore be described as "a description of the right angle—its scope, application and significance."

VIII Substance and form

The theory as it applies to nuclear particles

Having sufficiently prepared our tool kit, we are ready for further exploration. We have just described the priority of the experiential (threefold) world to the conceptual (fourfold) world, and the priority of both to the outer physical world of events, that unique "object" which we refer to as plainly and unquestionably *out there* in front of us, even though we often project upon it the substance of the threefold and the concepts of the fourfold.

Ordinarily, the fact that we project properties on objects needs only to be recognized when we make a mistake, as in a case of mistaken identity. The projection of substance, however, is an inevitable accompaniment to *all* perceptions, and the question of correctness does not properly apply. The quality of hardness, wetness, or solidity, even of goodness, beauty, or desirability, is supplied by an inner pressure that needs but the slightest excuse to throw itself upon the perceived object. Observe the projections of youthful love, which finds such desirable qualities in the object which it has singled out, as though it were only awaiting a container into which to pour what it itself has created. Or the play of children, where the crudest improvisations suffice to carry out the drama—a stick for a gun, a chair for a horse.

So the position we are taking, and calling upon the threefold operator to establish in formal fashion, is that this "substance" is a *prior condition,* supplied (perhaps by dreams or by imagination) *before* the lessons of life have trimmed and shaped it to conformity with actual circumstances. The latter operation is an enforced learning which, through the teaching of distinction and judgment, prunes the rank profligacy of projection and makes it possible for us to judge a situation objectively. Only in this way do we learn to distinguish, say, a bona fide antique from an imitation, or to make any wise purchase. We learn by mistakes and by careful application, even though the appetite for aesthetic experience is ready to leap at whatever moves.

Form versus substance in science

We are drawing up a scheme not just of psychology, but of the order of generation itself. So, to add greater scope to our inquiry, we should cast our nets in deeper waters.

In the physical sciences, we can find the same distinction between entities to which formulae apply (forms) and entities to which they do not (substance). But this distinction is not immediately apparent, for science is committed to the doctrine that substance is unnecessary. Locke first attacked the concept of substance by insisting that the real validity of knowledge is overthrown by the assumption of substance. Berkeley showed that the supposed material substance of things is not utilized either in science or in common life. Hocking, in paraphrasing Berkeley, states, "The chemist can always determine whether an object before him is gold, but he never does so by inspecting its 'substance.' He reaches his conclusion solely on the basis of its properties—its solubility in different acids, its combining proportions and weights. These are all he needs to work with. For is not the 'substance' of gold a mere name for the fact of experience that these properties belong together?"*

This is an excellent statement of the position of scientific objectivity, a position that has seen its most forceful exposition in logical positivism (only that which can be operationally tested is valid).**

But today, this reliance on relationship structure as the only truth with which science can deal is producing a rude shock in nuclear physics. As Gamow says:

> "Although experimental studies of these new particles reveal new and exciting facts about them almost every month, theoretical progress in understanding their properties is almost at a standstill. We do not know why they have the masses they do. We do not know why they transform into another the way they do. We do not know anything."***

This was written in 1959, but the situation has, if anything, become more confused. A number of new particles have been discovered, and there is talk that the true elementary particles may be much smaller.

*Hocking, William E., and Hocking, Richard B., eds. *Types of Philosophy* (p. 162). New York: Scribners, 1959.

**As another milepost in the history of this elevation of form as the objective and valid reality, we have relativity. Eddington says (in *Mathematical Theory of Relativity,* p. 9, Cambridge University Press, 1923), "If we describe the points of a plane figure by their rectangular coordinates, *xy,* the description of the figure is complete, but it is also more than complete because it specifies an arbitrary element, the orientation, which is irrelevant to the intrinsic properties of the figure and ought to be cast aside from a description of these properties . . ." Here, Eddington wants to toss out the orientation rather than the substance, but the statement illustrates well the vogue for form that relativity implemented.

***Gamow, G. *Scientific American,* vol. 201, no. 1, 1959. Compare Gamow's statement with Robert M. Eisberg's (in his *Fundamentals of Modern Physics,* New York: Wiley, 1961): "It is fair to say that the properties of the atom are completely understood . . . for the nucleus this is not the case."

The threefold operator and the nuclear realm

To me it is deeply significant that the nuclear particles have remained inexplicable, whereas atoms, once the key was found, have been completely explained. Atomic structure, mass, chemical properties, radiation spectra, etc., have all been accounted for and even predicted with great accuracy, but nuclear particles defy explanation by rational theory.

The reason, I believe, is that the realm of nuclear particles precedes form. It is the realm of raw substance: the possibility of form has not emerged, and will not emerge, until the atom exists. Form, which involves two dimensions of constraint, is conceptual and can be formulated, but that which precedes form, namely, substance, *cannot be formulated.*

However, it is appropriate to apply the threefold operation to nuclear particles. The nuclear realm introduces mass (rest mass) and motion (measurable and extrinsic motion, as distinct from the intrinsic and constant velocity of light). Both of these characteristics reflect the threefold. It also exhibits irreversibility (one-wayness) and other forms of asymmetry, such as parity (right- or left-handedness) and the fact that the proton has a much greater mass than the electron.

Further, and more significantly, the permanent nuclear particles, the proton and electron, do not possess identifiable structure as do atoms. This is because the principle of *form* does not exist prior to the atomic realm. Nuclear particles do not have form as do atoms, whose structure and behavior are so completely in accord with rational theory. Since the nuclear realm is prior to form, it cannot be accounted for in terms of form, and requires a different kind of description.

The threefold realm has two degrees of freedom, and since, as we said, forms or shapes require two degrees of constraint, they cannot exist at this level. In other words, a law here can deal only with linear measure; it cannot supply or impose a shape.

Like the bank, whose only requirement is that you may not overdraw your account, this "law" puts no stricture on how you "spend your money." Hence, we find with nuclear particles conservation laws of several sorts: conservation of mass, of energy, of momentum, etc. But we do not find laws like the Pauli exclusion principle, which dictates the configuration of electron shells in the atom.

This correlation with the threefold reveals a wider significance in the realm of nuclear particles. I believe that it may shed light upon cosmological and even metaphysical questions which science has heretofore felt outside its province, questions of ontogeny, or the order of generation from first cause.

This vague hunch on my part has prompted me to try to grasp something of the strange new world that now is being uncovered: the physics of nuclear and subnuclear particles. What could be the laws of this world?

I had a few additional hints from the threefold operation:

1. *Asymmetry.* According to our theory, we would expect nuclear particles to have one degree less symmetry than atoms. Since atoms have *radial* symmetry—a central axis with symmetrically disposed electron orbitals—nuclear particles would have the bilateral symmetry of a helix or screw. This has at least been indicated by the parity experiments of Lee and Yang. This discovery of asymmetry, disconcerting to physics, is a confirmation of the threefold operator, whose one-wayness is similar to that of a screw (right- or left-handed).

2. *Behavior.* Since nuclear particles have two degrees of freedom as against one for atoms, they should exhibit this difference in their *behavior.* (Behavior is a manifestation of freedom.) When nuclear particles radiate, or rather, dissipate

into radiation (for they do not remain particles), the disintegration products are other particles as well as radiation, adding thus an extra "dimension" to the product. Extremely interesting to me is that the difference in degrees of freedom manifests in a *literal* fashion, in that the radiation spectrum of *atoms,* displayed on a photography plate, is a one-dimensional array of frequencies:

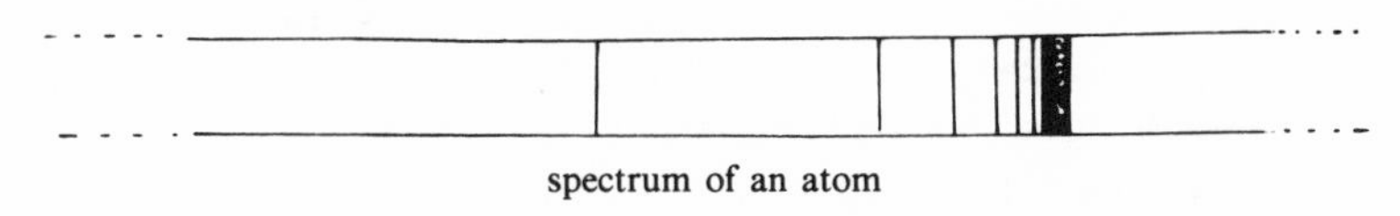

spectrum of an atom

whereas the radiation of nuclear particles displayed on a photographic plate is two-dimensional:

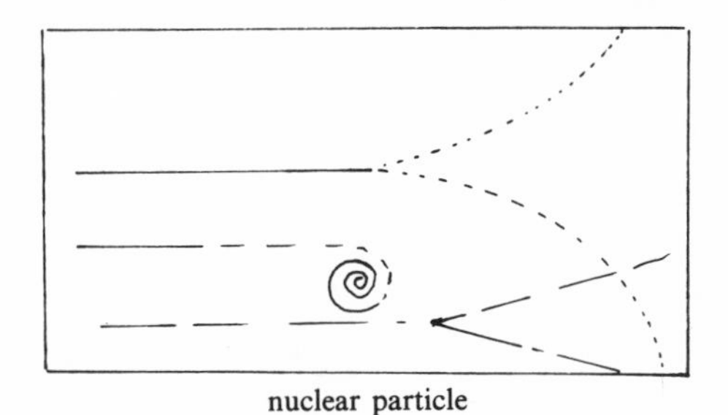

nuclear particle

3. An additional consideration that brings out the difference between the nuclear and atomic realms is of a rather abstract nature. It is so novel to our accepted notions that I'm happy to have found a statement Eddington made some forty years ago that can now be interpreted as anticipating this subject, the generation of nuclear particles. He says:

> "The fundamental basis of all things must presumably have *structure* and *substance.* We cannot describe substance, we can only give a name to it. Any attempt to do more than give it a name leads at once to an attribution of structure . . ."*

*Eddington, Arthur S. *The Mathematical Theory of Relativity.* London: Cambridge University Press, 1923 (p. 221 ff., in Section 98, "*General Relation Structure*").

He goes on to describe structure, and then continues:

> "There is a certain hiatus in the arguments of the relativity theory which has never been thoroughly explored. We refer all phenomena to a system of coordinates, but do not explain how a system of coordinates (a method of numbering events for identification) is to be found in the first instance. It may be asked, What does it matter how it is found, since the coordinate system is entirely arbitrary in the relativity theory? But the arbitrariness of the coordinate system is limited. We may apply any continuous transformation; but our theory does not contemplate a discontinuous transformation of coordinates *such as would correspond to a reshuffling of points of the continuum.** There is something corresponding to an order of enumeration of the points which we desire to preserve, when we limit the changes of coordinates to continuous transformations."

Now it has recently been discovered that nuclear particles are permutations, reshufflings, of several quite esoteric factors or parameters. Some of the parameters are little more than symbolic: spin, strangeness, charge, atomic mass. But these permutations produce a number of distinct entities: pi and kappa mesons, lambda, sigma, and xi hyperons, etc., all of which are transitory; i.e., they disintegrate almost immediately. Only two permutations are permanent: the proton and the electron.

Another recent development has been the notion of "quarks," elemental carriers of spin, isospin, and strangeness, whose *combinations in groups of three* account for the recognized particles with mass equal to or greater than that of the proton.

The fact that three quarks are necessary echos the threefold theme again. And recall our discussion at the end of Chapter IV, in which we recognized that mass was a threefold entity; and our discussion of cube roots, in which we saw that the

*My italics.

threefold can be expressed in unitary fashion only by using the three dimensions of space.

Eddington's first statement, that we cannot describe substance because "any attempt to do more than give it a name leads at once to attribution of structure," is essentially negative in that it says we cannot describe substance, but by this very acknowledgment it sets us on the right track and leads to the second statement, about a "reshuffling" of the continuum. Did Eddington connect the two thoughts? Whether he did or not, I would like to make the connection, and assert that substance correlates with the ordering of the points of continuum.*

I would further claim that recent developments in nuclear physics, in which the several kinds of particle are accounted for as permutations of some more basic entities, are a confirmation of this insight of Eddington's some twenty-five years before the complexity of the subnuclear world was known.

The fact that Eddington is talking about points in a continuum, while the entities of nuclear physics are not points but quarks or the like, does not affect the validity of the correlation. We must realize that "points" are *undefined entities.* And so are quarks. We cannot endow these elements themselves with spatial or even substantial nature, because they are only carriers of abstract properties. It is the *ordering* of quarks that produces what we recognize as mass or charge.

So much, then, for substance. We can well realize why the subject has not been pursued by science until recently, and why the present research of nuclear particles is encountering such difficulty. But while we do not have the gadgets of science, its synchrotrons, cyclotrons, etc., we are engaged simply as living creatures in a continuing interaction with our

*Manifest, perhaps, as the inversion of time in the world of antimatter.

environment, and thus have direct access to the substance of the universe. For what occurs in the cosmos at the nuclear level, the realm of substance (mass), the realm of asymmetry manifesting the threefold operator, corresponds to nothing other than experience itself. We feel, we fear, we hunger, we value. Even that universal conveyor of information, light, is taken into us, into our consciousness, consumed and converted, much as it is by plants through their photosynthesis. It becomes our understanding, coloring our imagination and lighting our dreams.

At the risk of seeming to wander far afield, I would like to end this chapter with a distinction between threefold and fourfold drawn from another area—human experience. This is the distinction between the emotional and the intellectual. A very good example of the former is the experience described by Carlos Castaneda in *The Teachings of Don Juan* and *A Separate Reality.*

By way of résumé, we may now say the encounter with reality consists of steps that follow and build on one another:

The first step is awareness itself.

The second is experience in time, a sequence of events accompanied by feelings. The second level—threefold: past, present, future—presents memory of past feelings and their relation in time.

The third step occurs when a present and a past experience are compared and a concept is found which measures or identifies an experience or what is common to experience. These concepts permit laws or generalizations to be formed about events, and the distinction between internal and external (objective) becomes possible.

The knower can now test these generalizations or laws.

IX Purposive intelligence and the twofold operator

I have described experience as the simple sequential ordering of events (threefold), and mind as the integration of these events into forms or concepts (fourfold). Often, however, our initial concepts prove inadequate or false, and the only way we can test them is by encounter with objective fact at the level of the tangible world. This is the domain of determinism, the realm of law.

All such laws may be characterized as dealing with cause and effect: they link two events in casual relation. And we must heed this law, for it is basic. If we do not listen to it, we only repeat our mistakes. However, once we are aware of its operation, we can invert our relation to the law and make it work for us. I hope to show this in what follows.

The law of cause and effect is the basis of the deterministic world view, which denies the existence of free will. The determinist, observing that everything in the universe has a prior cause, concludes that since man is part of the universe there is no ground for the existence of free will.

The geometry we have proposed, however, enables us to see the law of cause and effect in perspective. While the law manifests at the level of objective fact, its existence as a *concept* is, of course, at the fourfold level, and as *experience* at the threefold. And the knowledge of cause and effect must initially and ultimately depend upon the threefold—the experience of the temporal ordering of before and after. There is no way to describe cause and effect conceptually. We require experience to know which comes first.

But the determinist's "scientific" point of view takes into account only the manifestation and the concept; it does not include the experiential aspect. The statement, the concept, of cause and effect is the mapping of a relationship. It is given in spatial terms, in simultaneity, and time is factored out. The deterministic view, or the logical view, or, in fact, any "mapping" in conceptual terms, *omits temporal succession.* It can relate two events but cannot say which is first.

On the other hand, the world view according to our geometry does include a temporal succession. In fact, as stated so far, it imposes a fixed temporal succession, which would support the determinist's view. But now consider this example:

In the old story, the barn burns down and the pigs are roasted.

(1) Fire → (2) cooking. But then, the improved flavor of roast pig is discovered, and barns are burnt in order to cook pigs. Thus fires are made on purpose, and the purpose, to cook pigs, becomes the *cause* of the fire.

(1) Cooking (imagined) → (2) fire → (3) cooking (actual). This is where the fallacy of the determinist's view is revealed. The temporal ordering of *cause and effect has been reversed.* The cooking which was the effect of the fire has become its cause. A *purposive intelligence* has the power to reverse cause and effect.

In one sense, the reversal of cause and effect is a reversal of time. In another sense, it is not, for the envisioned result, the desired cooking, is imaginary, not actual cooking. So there is no "actual" or physical time reversal. The envisioned result or goal is nonphysical. Must we therefore dismiss it as nonexistent? No. For our cosmology postulates that experience, as threefold, is derived from some initial nonphysical unity. We can trace the envisioning of a goal, or *purpose,* back to the unity from which "physical" experience was derived. Purposive intelligence, taking the cause-and-effect relationship into account, realizes that a goal may be achieved by using this relationship. So it approaches experience from a different direction: it initiates action addressed toward a goal.

We have not yet found where this purposive intelligence arises. While the *intelligence* is at least partially dependent upon the faculty of comparison, and would thus be fourfold, *purposiveness* is neither in the fourfold nor in the threefold. It is certainly not in the objective world, and so there is only one place left for it: the initial unity from which both the fourfold and threefold are derived by division.

And this makes sense. Recall that we have referred to the initial unity as a projective potency. This concurs with the projectiveness of purpose and its power to initiate action leading to goals. Here, we might note in passing, we have confirmed teleology by uncovering purpose at this fundamental level of the universe.

There are good reasons for having said little about this initial unity itself. Since it includes all qualifications at once, no qualifying words can be used to describe it—let alone "define" it. For definition doesn't come into existence until the level of the fourfold division.

All we can say, then, of this initial and unitary totality is that it has dynamic and projective potential. This potential, by declension or division of itself, produces that which can be defined, but not without a residue of that which cannot be defined. In thus containing contraries, the initial unity is beyond knowing in the usual sense. At the same time, it is the source of knowing and of being, and, of course, all the qualifications we have discussed under threefold and fourfold divisions.

But what does the initial unity have to do with cause and effect? Simply this: it is the place of the *first* cause in any causal sequence. It is the *purpose* which initiates a chain of causal relationships.

Introduction of the twofold operator

Return now to our first discussion of the learning cycle. While we found that we could use the same diagram to describe both the learning cycle and the controlled motion of a body, recall that the *order* in which the steps were taken had to be reversed. While the analysis of motion used clockwise rotation on the diagram, learning was counterclockwise.

Having now distinguished two kinds of action (action toward unity, *toward goals,* and action away from unity, *toward manifestation*),* we can recognize their equivalence to the action which moves clockwise on the diagram and the action which moves counterclockwise.

The latter motion is toward means, toward manifestation; the former is toward goals. In the learning cycle, we have instinctive motion which causes encounter with the unexpected (the child burns his finger) and results in learning and self-control. In the *application* of the cycle of action, or the control of means to achieve ends, the rotation is in the opposite direction. Observation first, then reaction to produce what is wanted. Generalizing, we may say that clockwise rotation is *informed* or *premeditated,* whereas counterclockwise motion is natural or naïve.**

*By manifestation I mean any step in the direction of making known to the senses, toward making tangible, visible, i.e., toward *effects.*

**Recall that, moving clockwise, each step in the cycle of action was the derivative of the one before. Since the derivative is equivalent to a ratio or slope, it is *predictive.* This is another sense in which the clockwise rotation is the *informed* direction.

As yet, we have no convention in our theoretical scheme for distinguishing naïve action from informed action. But these considerations are so basic that for their formal expression we need a new operator—a *twofold* operator—to accommodate left- and right-hand rotation in our system.

It is important to realize at the outset that this operator is not to be confused with "pseudo-dichotomies" such as positive and negative, true and false. Such apparent dichotomies, as I explained, actually involve four distinctions. In the case of positive and negative, the two hidden distinctions are positive and negative imaginaries; in the case of true and false, the distinctions are consistent and inconsistent, without which *no use could be made* of the true–false distinction.

The introduction of such an important idea as the twofold operator at this late stage challenges us to do some very basic thinking. Why have we not considered it before?

Let us first note that it is much easier to talk about the divided than the undivided. We began with the fourfold division, because it was the most accessible to rational treatment. We then took up the threefold only to explain that it deals with an area (experience) that is inaccessible to definition. Of the initial unity, we can say nothing, except to regard it as a dynamic potency (we cannot predict where such potency will lead).

It is natural, therefore, that the twofold should be inherently subtle and difficult to grasp. But we have found that it comes about through a change of sequential ordering. Let me use a concrete example:

An electric motor connected to a source of electricity may be used to pump water uphill. By reversing the sense, water flowing downhill drives a turbine (a pump in reverse) connected to a dynamo (a motor in reverse), producing electricity. A diagram of the two arrangements would be the same whether the electricity was consumed to pump the water or the water used to produce the electricity. There is no logical or analytical distinction between the two. To make the distinction, we have to talk about the *order* in which the elements are taken—which is input and which output?

This example shows that the distinguishing mark between the two situations—or between what we are calling right- and left-hand rotation—is the sequence of the three elements *a, b,* and *c.** This distinguishing mark can be perceived only as a difference in temporal succession.

Considering cause and effect, then, we can no longer say that these are twofold, because there is a third factor describing which comes first.

The twofold operator in nature

We may now ask an important question. Why does the twofold operator not come into play before the threefold and the fourfold? It would seem reasonable to have the twofold come in between unity and the threefold. Why does it not come into play until experience (the threefold) has been analyzed by intelligence (the fourfold)?

The answer is that it does. At each stage after unity, the twofold is operating, but of the two orders (or motions) *only one manifests.* This is the "natural" order, which invests in substance and form. It moves toward multiplicity and away from unity. The other order of the twofold moves away from substance and form toward unity (the goal). But if this latter order or motion occurs too early, before substance and form are attained, the movement turns back from manifestation toward the unmanifest, back toward the unity from which

*Three elements are required to establish an order; (. . . *ababa* . . .) does not define an order, whereas (. . . *abcabca* . . .) defines an order which is different from (. . . *acbacba* . . .).

manifestation is derived by division. And we do have evidence for this turning back at the earlier stages.

There is a phenomenon in the production of nuclear particles which I interpret as an example of this clockwise rotation at the second stage of cosmic process. A very high-energy photon (pulse of light) has the capacity to turn into a material particle (an electron or a proton). When this occurs, there is simultaneously produced an antiparticle (a positron or an antiproton). The antiparticle almost immediately reverts into radiation or photons. Antiparticles are called antimatter, and we may interpret their production and subsequent disappearance as an example of the clockwise rotation at the level of nuclear particles, noting their nonproductiveness at this stage.

I have recently done some speculation on antimatter. We know that the mass (i.e., energy) of antimatter cannot be negative, because when the photon produces a particle and an antiparticle, the positive energy of the photon is equally divided. The reversal must be one not of mass but of *time.* I prefer to think of the reversal of time not as *negative* time ($-T$) but as inverse time ($1/T$). This is, of course, conjectural, but it has some interesting implications. If one thinks of normal time as being very long (even if not infinite), then inverse time ($1/T$) would be very short—eternity in an instant. In the photon, it has long been known that the energy is inversely proportional to time ($h = ET$). This implies that in the "anti" world there might be an unlimited amount of energy in an instant of time, reversing our normal relationship between size and importance.

This compaction of time would give it the character of *omnipresence*—not going "backward" in time, away from the present, but instead going more deeply into the present. This interpretation has the merit of conforming to references in countless religions and mythologies to the super-sensible, nonphysical celestial world. "The Kingdom of Heaven is like a mustard seed . . ."

Returning to the operation of the twofold at subsequent stages of cosmic process, we find a corresponding duality at the third stage, the generation of atoms. There are certain kinds of atoms, helium, argon, neon, etc., known as noble gases, which have complete electron shells and do not enter into chemical combination with other atoms to form molecules. This "refusal to manifest" (by engagement with other atoms) can be interpreted as the clockwise rotation again, and, like antimatter, is nonproductive.

We have already, in Chapter VII, shown the progressive loss of freedom at successive levels.

Level		
I (Light)	Unity	Three dimensions free
II (Nuclear)	Threefold	Two dimensions free
III (Atoms)	Fourfold	One dimension free
IV (Molecules)	Twelvefold	No dimensions free

This declension, however, depicts only the process of generation produced by the counterclockwise rotation, toward manifestation. The clockwise rotation, away from manifestation, is of course toward goals; but it is also toward greater *freedom* (more degrees of freedom, fewer degrees of limitation).

At the fourth, or molecular, level, a reversal can occur in the evolutionary process, and the appropriate direction moves now toward goals (what we have called clockwise rotation). Further counterclockwise rotation results only in molar aggregates, which do not participate in the process. The molar aggregates do not comprise another level because they are not fundamentally different from molecules in terms of the theory. As aggregates, they have lost that trace of the original potency which appeared in individuals on the molecular level as the phase dimension, or choice of timing. They constitute the inert "furniture" which is used by process to fulfill itself.

We can now realize the reason for the nonproductiveness of the clockwise rotation at levels II and III. When the clockwise rotation occurs at the nuclear and the atomic levels, before the complete loss of freedom at the molecular level, it cannot harness means (matter and law) to its purposes. The opportunity to use means comes at the fourth level, in the physical world. This is the world of commitment, where the initial freedoms have changed into restraints; the world of determinism.

This world is characterized by the increase of entropy. Entropy is the tendency of differences to average out, of stones to roll down from the mountains and fill valleys. In other words, it is the tendency for energy to become more uniformly distributed and hence unavailable.

In the molecular world, the reversal of direction permitted by the twofold operator does have significance, and a crucial one. It makes possible the *decrease* of entropy.* A small system can store order or energy which it draws from its environment, and build this order into itself. Such a system is the living plant.

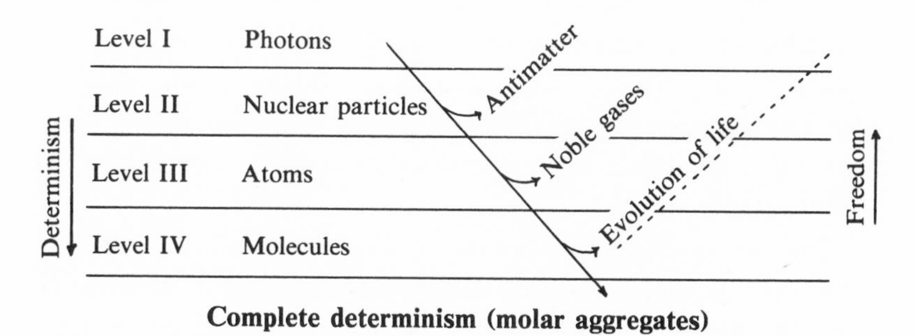

Complete determinism (molar aggregates)

*The decrease, that is, of *local entropy.* While the total entropy of a particular system always increases, it is possible in living systems to circumscribe a *local* region where entropy decreases; in inert systems no such region exists.

The emergence of life

Living things have the power to organize inert matter into complex and self-perpetuating forms. In human activity, this would correspond to the employment of law to produce an effect—using the fire to cause cooking. The demarcation between the living and the nonliving may be variously drawn, but the crucial distinction hinges on this capacity to oppose the increase of entropy, to store (and expend) order. The reversal of entropy does not imply a mere inversion of the former tendency toward disorder. Life process involves both conservation and expenditure of energy, but requires a net gain, just as a business must make more than it spends.

This view of the initiation of life provides a deductive and theoretical basis for what has hitherto remained a paradox, an apparent violation of the laws of matter. We mentioned earlier that the conclusion of determinism (there can be no free will) is shown to be erroneous when it is realized that the order of cause and effect can, in the human case, be reversed, and a desired effect can be the cause which is to produce itself.

We now extend this concept to apply generally. We observe in complex molecules the capacity to store order. There is no violation of the conservation of energy here; the order or energy is drawn from the environment. There is no creation of energy; the energy is made available by judicious sorting and storing in available form (as starch or sugar is stored in plants, for example).

Strictly speaking, we cannot, and do not have to, explain *why* this "judicious sorting" comes about. Since it involves the twofold operator, which carries back to the initial unity, we can appreciate that, as with all first causes, it cannot by definition be accounted for in terms of antecedents.

Nevertheless, the plausibility of our thesis is supported by relativity and quantum physics, both of which show that individual entities, in addition to physical volume, have an extra dimension which is equivalent to a *free choice of timing.* This is called the phase dimension and has the measure 2π.* It supplies the necessary condition for reversal of entropy, making it possible for an individual entity to store order. These two different (but in the last analysis, equivalent) arguments run roughly as follows:

1. *The argument from relativity.* The volume of the universe according to Einstein and Eddington is $2\pi^2 R^3$, not, as it would be for a physical sphere, $(4/3)\pi R^3$. The difference, as Eddington explains it,** is that the "ordinary" sphere, $(4/3)\pi R^3$, is multiplied by two added factors. One of these is 2π, which Eddington calls the phase dimension. This phase dimension is an *uncertainty of direction of* 2π, it being the character of uncertainty that its measure is angular. Thus we can describe the defining power (or certainty) of a lens as the angle subtended by two points that it can just discriminate. The smaller this angle, the greater the definition of the lens. It follows that the maximum uncertainty (lack of definition) is the largest possible angle. This is not infinity, but 2π, the whole circle. (As the *whole* circle, this is also a unity.)

A second factor, 3/4, comes in as what Eddington calls "fixing the scale," which we may interpret as a reduction resulting from self-limitation. (See Chapter II: 3/4 point = control.) Eddington makes a provocative comment on the significance of this stabilization of scale: "But now that each particle, or small system has its own scale variate, a new field

*Here 2π means an angle of 360 degrees—it is called phase, and is a dimension because it is an additional unknown. For example, we say, "I don't know how this will turn out," or, "I don't know which way he will take this." "Which way" implies he may be for it, against it, or indifferent—this range of possibility can be covered by the angles possible in a circle.

**Eddington, Arthur S. *Fundamental Theory* (p. 47). Cambridge University Press, 1946.

of phenomena is opened up to theoretical investigation, which is suppressed in the molar treatment of scale as an averaged characteristic."* The phenomena suppressed in the molar treatment are made possible by the 2π phase dimension and the 3/4 stabilization of scale. These I interpret as:

a. Capacity to make choice of timing or 2π.**
b. Self-limitation or stabilization of scale, 3/4.

The two factors, 3/4 and 2π, which multiply the ordinary sphere, $(4/3)\pi R^3$:

$$3/4 \times 2\pi \times (4/3)\pi R^3 = 2\pi^2 R^3$$

thus render any small system, volume $(4/3)\pi R^3$, dimensionally equivalent to a universe (dimensionally equivalent means having the same essential nature, not the same size).

2. *The argument from quantum theory.* According to Planck's discovery, the photon, or pulse of electromagnetic radiation, is described as a quantum of action:

$$\text{Photon} = \hbar \text{ (Planck's constant)}$$

This quantum of action is also described as a quantum of uncertainty and contains the angle 2π, a phase dimension, just as in the formula discussed above.

In other words, the light pulse is a "piece" of uncertainty. I cannot make this more palatable except to say that it can only be this way. Were it possible to predict or define or account for this uncertainty, we would simply transfer the ultimate reference, the first cause, to something else. "Where do babies come from?" "The stork brings them." "Where does the stork get them?" etc.

* *Ibid.*

**There may seem to be a distinction between 2π as direction (in space) and "timing." But all time indicators in a finite universe have to be cyclic and as such show time as spatial angle.

What is interesting and possible to account for is the chain of effects which this photon can produce. If it is very powerful (the smaller its wavelength, the more energy it packs), it can become a nuclear particle, either a proton or an electron. Some of the uncertainty has become mass (or certainty), but there is still retained a proportion of energy which is in the form of radiation (or uncertainty).

A further step combines the nuclear particle and other particles into an atom with further and very considerable loss of uncertainty, followed by still more at the molecular stage. At this molecular stage the remaining uncertainty, now much reduced in energy, persists as the bond which holds the molecule together. (We may think of a bond as equivalent to the energy necessary to break the bond.)

The measure of this uncertainty still contains 2π, the choice of direction or phase, despite the fact that the *quantity* of energy is reduced. (As an example, think of a person with a much smaller pistol, but still having free choice of direction to shoot.) With the development of large molecules, this bond energy gets smaller and smaller until a point is reached where the energy of the bond and energy of a random molecule in the environment are of the same order.

At this point it would be possible in theory for the molecule to act as an energy sink, to draw energy from the environment and store it as order.

Now, the fact that life is very sensitive to temperature, that refrigerators preserve food because they prevent the chemical action of bacteria and molds, that fever and pasteurization destroy germs, and that harmful bacteria cannot exist at perpetual temperatures over 110 degrees Fahrenheit, confirms this speculation about bonds. The bonds of molecules involved in life processes must involve quite definite energies with rather narrow limits (plus or minus five percent in general, plus or minus one percent for warm-blooded animals). The energy of these bonds is of the same order as the residual uncertainty—about one-twentieth of an electron volt.

We see then that the theory of relativity and quantum theory agree. Both indicate that in large molecules there is a modicum of energy and choice of timing (phase dimension, 2π) due to which the molecule can extract energy from its environment and build order or organization. Within a certain temperature range this emerges as life.

The evolution of life—increase of freedom

Up to this point, our account has described how the physical world of objects has come into existence. In contrast with the more "natural" explanation, which accepts physical objects as "being there in the first place," our interpretation holds that the originating and primary "stuff" is unitary, undifferentiated, devoid of properties,* without form or substance; it is known to physicists as radiation.

Out of this primordial unity, there emerges, by division of itself in threefold fashion, the flow of time.** This division establishes the possibility of motion, of substance, and of relation. By a further division of itself, the fourfold, space (as a matrix of relationship and independent of time) comes into existence. In human terms, this fourfold division makes concepts possible; in terms of generation, it initiates form and interrelationship. From the combination of these two divisions, the threefold and the fourfold, comes the generation of the physical world, the world of inert molar objects, subject to the laws of determinism.

*Wavelength or frequency in our view is not so much a property as a potency of the photon. The velocity of light is not so much a property (i.e., limitation) of light as it is a limitation on matter.

**Time does not exist for radiation, because at the speed of light clocks would stop.

According to this view, matter, or the world of inert objects subject to law, is not fundamental, but derived, and is three levels removed from the primordial and undifferentiated unity. Similarly, determinism, whose laws render inert objects predictable, is a derived principle, and owes its existence to the canceling out of the individual uncertainties of particular particles or subsystems.

However, the advent of determinism is, as we said, the last of a series of stages by which the initial freedom of the undifferentiated substratum (electromagnetic radiation) spent itself, first to acquire substance, then form, and finally concrete objective existence—a stake, as it were, a participation in the unique manifestation which we call the "physical world."

But the space–time of the physical universe is singular, unique. (As Kant pointed out, the remarkable property of the physical world is that there is no other. Heraclitus also noted the uniqueness of the physical world; he said that in waking life all men share one world, but in sleep they create their own and different worlds.) This fact, actually a corollary of the principle of determinism, makes it a truism that two bodies cannot be in the same space at the same time, that airplanes crash if their flightpaths cross. In short, it makes collisions possible, but by the same token provides laws which can ensure against them.

The progress of generation has carried us from the extreme of an initial and undifferentiated dynamic, to the contrary extreme of a derived and diversified objectivity (an unlimited number of kinds of molecule). The one is completely free, the other completely determined. The "descent" can go no further.

The ground is prepared for the next and only possible step, the reflection, the rebound, of what remains of the primary dynamic from the inflexible and unyielding barrier presented by the world. This possibility of a rebound is provided by the free phase dimension described above. The appropriate action, by foreseeing the result, can create the cause that will produce the effect. The *result* is the reversal of the tendency toward greater entropy, the possibility of drawing energy from the environment and storing it as order. This is the first step toward the regaining of the freedom lost, the fifth step from the beginning.

We are now ready to explore the nature of this first "upward" step and then the other possible steps. We have a choice whether or not to make use of empirical references to show how these steps manifest in nature. It would be easier from the point of view of clarity to do so, but the clarity gained would be at the expense of our deductive method. We are trying to lay out the course of evolution from first principles by dead reckoning, "flying on instruments," and we would be going against the spirit of the endeavor if we "look out the window" to see where we are.

So I will ask the reader to bear with me and concentrate on deductive principles. These principles have already been established; we have only to draw out their implications.

First we recall how the threefold and then the fourfold operators brought about the loss of the freedom intrinsic to the original photon. The threefold created the limitation of movement along a time line which was one-dimensional, sequential, and nonending. Then, with the fourfold operator, there was limitation in two dimensions. This produced forms and shapes which are necessarily finite, so that the first limitation, a commitment to one dimension with a freedom in two, was exchanged for a freedom in one dimension and a commitment in two.

If freedom is regained in the same way that it was lost, we may expect the first step toward freedom to conquer time, because time was the first constraint, and the second step to conquer space, because space was the constraint to which atoms are subject. Our first freedom will be infinite but one-dimensional, and the second finite but two-dimensional.

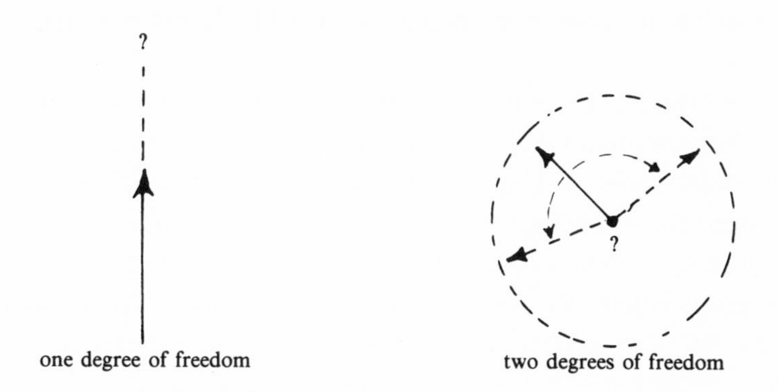

Finally, there will be the sum of these two; that is, unlimited goals in any direction.

Thus we may anticipate three distinct kinds of entity:

1. Entities moving toward an infinite goal with no choice of direction (one degree of freedom), conquering time.
2. Entities moving toward finite goals with choice of direction (two degrees of freedom), conquering space.
3. Entities moving toward a nonfinite goal with choice of direction (three degrees of freedom).

Categories of life exemplifying the three degrees of freedom

To the first class of entity we may assign *plants* (the vegetable kingdom), on the grounds that both growth and self-propagation comply with the definition "moving toward an infinite goal with no choice of direction." Heliotropism (growth toward light) is a specific behavior of this sort. Propagation through seeds conquers "the dimension of time."

To the second class of entity we may assign *animals,* since the mobility (voluntary motion) of animals, their pursuit of food and of each other, complies with the definition "moving toward finite goals with choice of direction" and conquers space.

The third class may be assigned to man, on the basis of his greater command of his environment, his greater freedom. This increased freedom is demonstrated by the pursuit of goals beyond his immediate needs and capabilities. Since the beginning of history, man has been motivated by principles, religious beliefs, or even by the pursuit of power for its own sake. He also exhibits this greater freedom in his manufacture and use of tools, his creation of language and culture, his pursuit of the arts, all activities which indicate a principle different from that underlying the voluntary motion of animals in pursuit of food.

There are reasons for reservation, however—not because man is a species of animal, for he may well also be something else, but because we should expect to find at this level a whole spectrum of development, a correspondence to the range between the lowest and highest forms of animals, between amoebae and elephants, in short a development that leads to supermen and to gods.

In this anticipated spectrum, man as we know him is by no means the highest development. But the principle here is quite independent of shape. Though we can recognize a great difference between one man and another, one is barely able to learn simple manual skills while another composes symphonies, both might look alike. Therefore we should not expect the principle underlying this third level to manifest in a variety of external shapes as did the animal principle.

We will not go into the development of plants, animals, and man here; it is the subject of another book by the author, *The Reflexive Universe.** We need note only what is implied by the

*New York: Delacorte Press/Seymour Lawrence, 1976.

development in general: the freedoms which were lost in the descent into matter are regained in the ascent of evolution, but with a significant difference. What was random and blind on the descent becomes voluntary or controllable at will in the ascent. The trap of time and the trap of space become the power to organize forces and to animate forms. The means that brought the fall makes possible the ascent, and what is at first a constraint, when understood, becomes the agent for freedom.

X Applications

Let us now give some attention to the application and significance of the formal system we have established, based on three operators: a fourfold, a threefold, and a twofold.

The fourfold operator

The fourfold operator applies to analysis, concepts, forms. Consisting of two dichotomies that are mutually orthogonal, or independent, it tells us that in any situation that can be dichotomized, or divided into two opposite aspects, there must exist a second dichotomy which mediates or measures the first.

Consider the old story of Solomon and the two women. Both claimed possession of the baby, so Solomon proposed that the baby be cut in two and each woman given half. One woman agreed. The other said no: she would rather give up possession of the baby and have it live. Solomon recognized the latter as the true mother and gave her the baby.

What happened? Solomon split the dichotomy of possession by introducing an independent dichotomy: the *being* of the baby as opposed to its nonbeing. Under this scrutiny, like that of a painting under x-ray, the false mother was revealed.

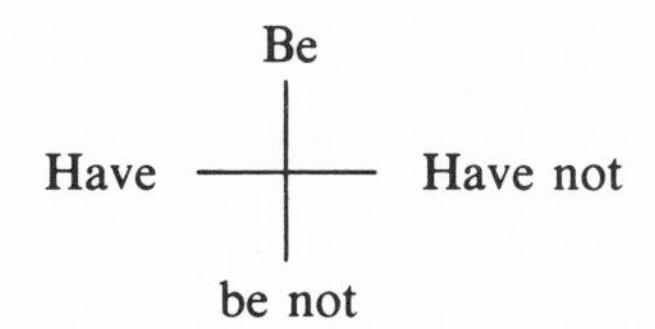

Problem: A married couple quarrels. The wife wants to go into a field, the husband insists that she must not because there is a dangerous bull in the field. The wife says the bull is not dangerous.

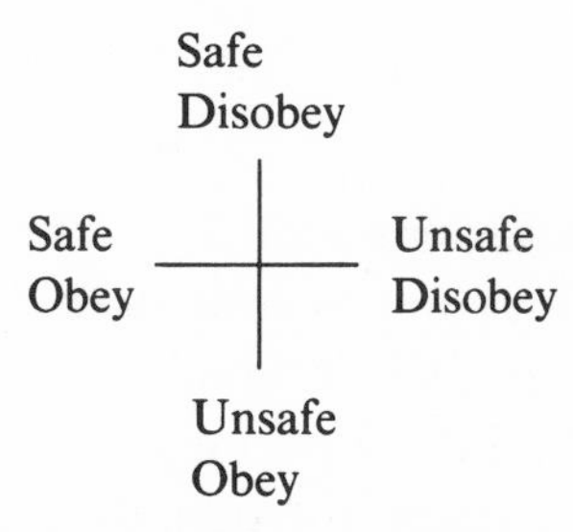

If we recognize that the situation contains two issues, safety and obedience, we can combine them differently: We ask the wife if she would obey her husband if the bull were really dangerous. She says no. We ask the husband if he would insist on the wife's obedience even if the bull were not dangerous. He says yes. Obviously, the couple is arguing about obedience, not about the bull.

Generally speaking, we try to discover two independent issues between the two parties. We describe the claims of the two parties in *two* opposite word pairs.

Once this is done, the rest is easy. We take the two word pairs and cross-product them. Recall the learning cycle in which the infant reaches out and touches things (an example which dealt with kinds of action). In that case, we were able to describe step 1 as *unconscious action* and step 3 as *conscious*

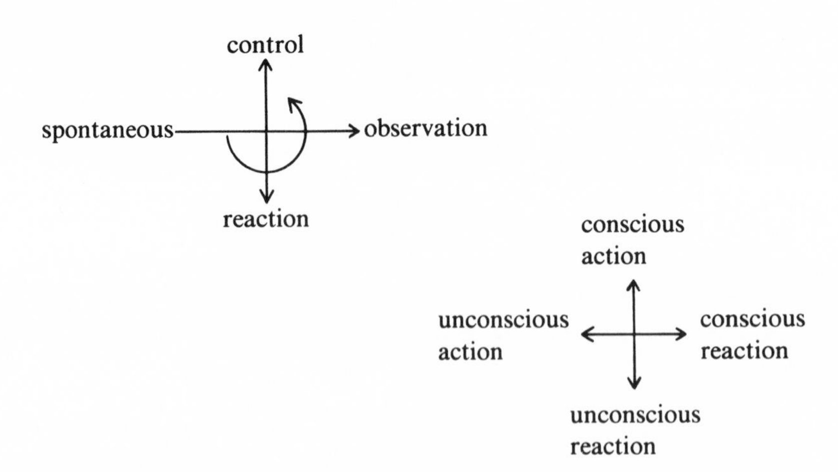

reaction (both words opposite). The other pair then become (2) *unconscious reaction* and (4) *conscious action.*

The same principle can be applied to kinds of relation, or kinds of "things." Take, for example, the contributing factors in a clock (relating them to Aristotle's causes):

Factors	*Aristotle*
A clock's function, to keep time	Final cause
The material of which it is made, brass	Material cause
The plan to which the parts conform, the blueprint	Formal cause
The work of putting it together	Efficient cause

Now let us fit these causes into "pairs of opposites":

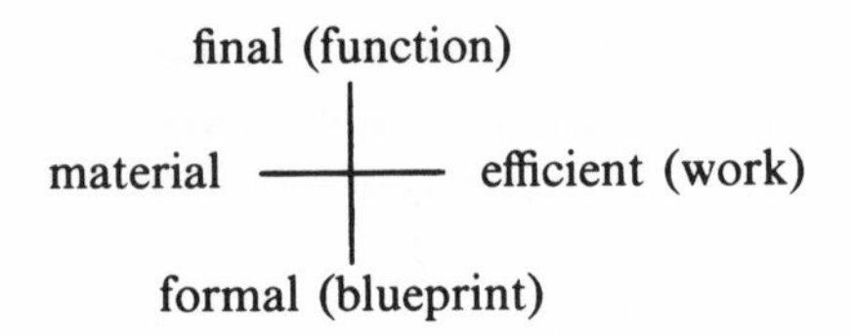

Obviously, the function is projective. The blueprints are objective. Since the work of assembly is also objective, we are forced to say the material is projective. This may seem untrue until one realizes that the brass could be used for many things besides the clock. This indicates also that the material is general; hence, the work is particular.

function {projective, particular}

material {projective, general} — work {objective, particular}

plan {objective, general}

We have to choose whether the plan is particular or general. It has to be general because any number of clocks can be made from it. This leaves us with timekeeping, the function, as particular. Here we must recognize that timekeeping, while it seems a general function, actually is not. That is why one can ask "What time is it?" and expect a particular answer. If the watch kept general time, it would not mean anything at all.

The very difficulty of this example tells us what to look for in others. We must overcome the scientific habit of confining attention to what is true generally and objectively, while omitting what is true in the particular situation. Of course, an observed fact is always particular, and it is often very important to give attention to such facts. But this warning has often been stated. What is difficult is the projective particular, the existence in every situation of a "pointingness" that is not general. Like the question "What time is it?" it has one and only one answer, and this answer varies from moment to moment.

We have already used the map example, but it would pay to

note once more the great difference between the kinds of information that fall under each of the map's four aspects. The one we are most accustomed to, because it is intellectual, is the map as a statement of relationship between geographical points. From the map we can see that Albany is between Buffalo and Boston, and so on. We cannot know how long it will take to get from one place to another unless we have additional information, the *scale* (under which we include the mode of travel, whether we walk, ride, fly, etc.). This information is not on the map. It involves reality. We need to draw on experience. Distances in New York State and distances in the Himalaya Mountains might have the same appearance on a map, but involve altogether different orders of difficulty in their actual traverse.

Here we must not be misled by the fact that scale can be stated as a ratio on the map, because the ratio does not supply the missing ingredient, which must come from experience. On the map it says "10 miles to the inch," but we must know what a mile is. We cannot get this from a map. We might have walked a mile to school when we were young, or run a mile in five minutes on the track team, but our experience of a mile did not come from the study of geography books. It is of a different nature than the conceptual grasp of relationship.

Next, "Where are we on the map?" This is particular and objective: particular because it is a present fact, and objective because it can be communicated to others. This gives no difficulty because, since it is objective, it is given recognition by science, though logic might not accept it because logic gives no formal recognition to facts as such.

Finally, we come to the *particular projective.* "Which way to Detroit?" (We need know only one orientation. This, together with the map, will provide any other orientation.) But this is particular and projective: particular because it changes as we move about, and projective because it literally projects. It doesn't tell us a *present* fact, but a *future* fact. Hence, it is *projective.*

Note that we are talking in every case about the whole situation. The map is dealing with information about the terrain, to be sure, and may include a great deal of information about the location of towns, roads, etc., and a scale of distances. The significance of the fourfold information is that it goes further: it also includes, or has room for, *your participation in the situation,* how much effort and time will be required of you, and by its inclusion of the compass, makes it possible to include an ultimate goal, which objective science has to ignore.

The threefold operator

The threefold operation is an entirely different way of cutting the cake. It is much more fundamental, and it cannot be analyzed.

For this reason, it is difficult to write or talk about, although from the point of view of philosophy or of life, the threefold is fundamental.

In fact, the threefold is the natural way we move in life. We see something, buy it, and enjoy it: food, eating, satisfaction; stimulus, response, result. There are three categories of terms: relations, acts, and states. The reader will recall the difficulty of finding sufficiently general words.

To implement description, it is helpful to correlate the three categories to past, present, and future. The relation category correlates to the past because it is generally that which already is. The action category is in the present because all action is in the present. You are *now reading* this book. The state or affective category applies in general to the future in the sense that it is the result of action.

But it is also possible for the state to come first. I'm hungry, so I look for food. The state leads to the action.

Perhaps what makes the threefold so elusive is that whatever is named becomes a "relation" term: "Would you like to play tennis?" The action here is not playing tennis, but the question "would you like." "To play tennis" is held up as an object. This takes us into the complexities of grammar, for we must distinguish the principal verb from the infinitive phrase.

In fact, despite the risk of added complications, it is of interest to compare the basic structure of the sentence with the threefold, for the sentence has three principal parts: subject, verb, and object. "The hunter shoots the bird." The similarity is greatest in that the verb contains the action which produces a result. To press the similarity further creates problems, for in the sentence, A acts on B, one thing acts on another, whereas the threefold would make the action of the hunter transform the bird into a museum specimen. The threefold seems to deal in a wider range of significance. We are interested in what the hunter proposes to do with the bird—what is the result? Without this information, the threefold is incomplete.

Another characteristic of the threefold is that it is so natural that it blends into life; it does not "stand out." Indeed, we might suspect that there is a tendency, in the matter of philosophic norms, for the empty wagon to make the most noise. Dr. Johnson kicked the stone to prove that it existed, that it was objective, but both the word "object" and the word "exist" denote something that stands out or against. Against what? Since the word "object" derives from the Latin *ob,* "against," and *jacere,* "to throw," there must be something for the object to oppose, and this "throw" or projection is as much a part of reality as is the object. The threefold is this ongoing movement from one state to another, which takes place in time, or we can even say *is* time. In any case, it is experience and contains situations, actions, and results.

The fact that only the relation part of this can be communicated and defined leads to the quite erroneous insistence by philosophers on objectivity as a criterion of reality. Scientists, however, whatever they may say about objectivity, use all twelve measure formulae. But the measure formulae cannot all be objective. If we use tangibility as a criterion of reality, acceleration, which is immediately felt, is real, velocity is not. If we use visibility as a criterion, position is visible, velocity and acceleration are not, and so on.

Psychologists and teachers are discovering likewise that animals or children cannot learn without doing. They find that the learning process does not consist of filling a creature with information, as one would program a computer, but of active interaction with the environment.

But in these instances, the scientists are still seeing only from the outside. As manipulators, they are not experiencing the terms they manipulate. To talk about the force of attraction is quite different from to be hungry or to be in love. The scientist speaks of gravitational force and nuclear force with the same aplomb despite the fact that the latter is 10^{39} times greater. Why not? Well, just to display this difference on the same graph would require that we compare the diameter of the universe (10^{26} centimeters) to its smallest possible distance, the diameter of a proton (10^{-13} centimeter).

It is customary to think of love, attraction, pleasure, pain, etc., as subjective, not "out there" in the universe. Hume is immortalized in histories of philosophy because he showed that we do not know causality objectively. It is a mental habit. The history of Western thought has tended first to divide the universe into interior and exterior, and then to discredit the interior by saying it is not there, or is purely "subjective," by which is meant interior to persons.

What is overlooked in this reasoning is that the universe also includes the nonobjective factor. The proton attracts the electron; the nuclear force binds the atom; the planets move in closed orbits; the *universe has feelings.* Physicists call them forces. To deny their existence in favor of objective "formulations" is pure sophistry.

In sum, we have shown that the threefold, despite the seamless quality which makes it difficult to get "hold of," is still implicit in the measure formulae of science. These expressions, which constitute a basic vocabulary of science, fall into three groups of four.*

Actions	*States*	*Relations*
Position L	Moment ML	Power ML^2/T^3
Velocity L/T	Momentum ML/T	Inertia ML^2
Acceleration L/T^2	Force ML/T^2	Action ML^2/T
Control L/T^3	Mass control ML/T^3	Work ML^2/T^2

The rosetta stone

Physical quantities correlated with their equivalent English meanings

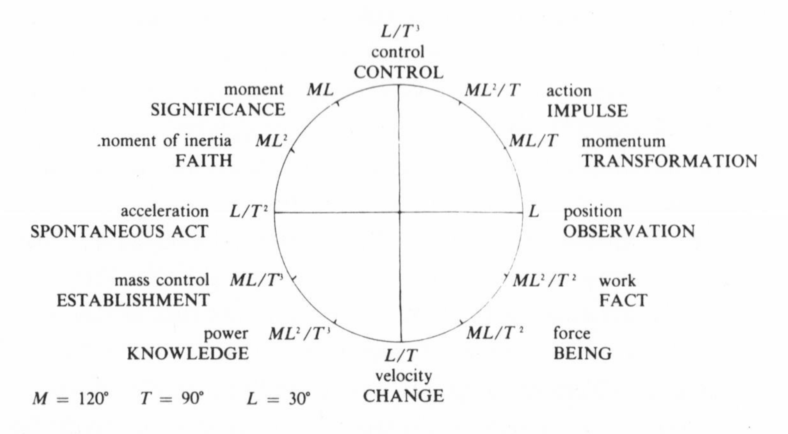

We need not review here the English meanings we are able to attach (see "Rosetta Stone" diagram). The point we wish to

*The last column is displaced one place from the tabulation in Chapter IV in order to have the three members on each line 120 degrees apart.

emphasize is that the measure formulae bear witness to the important status of the threefold in science.

There remains an important point that has not been covered. We have mentioned that the threefold can be correlated with past, present, and future. But we could go further, for the threefold may be even more basic than time, in the sense that it gives to time not only its customary division (past, present, and future), but its *directionality.*

The direction of time, like other aspects of the threefold, is not objective, and it cannot be communicated as we communicate comparative measures (by drawings, definitions, etc.). Immersed in time's flow, we have no opportunity to get outside it. And the threefold involves this directionality. If represented by three letters *a, b, c,* we can have two possibilities, *abc* and *acb.* Analytically, we can say that *a* is between *b* and *c,* but we have no way to say that *a* is *before b* except by reference to "subjective" time.

The choice of which way to go, *abc* or *acb* (clockwise or counterclockwise), takes us to the twofold operator.

The twofold operator

This is the most difficult operator. It is the one that gives the threefold its direction. We can have *only two such directions,* and this is why the twofold is different from one of the dichotomies in the fourfold. We might have thought, for example, that true and false is a dichotomy, but the teaching here is that it is not. It is *one of two pairs of opposites* that belong together in a fourfold. Thus we would have:
And note that this is just how it occurs in fact. We don't ever have a situation in which true and false suffice to describe it.

Take, for instance, a lawyer cross-examining a witness. He says, "Where were you on the night of . . . ," etc., and

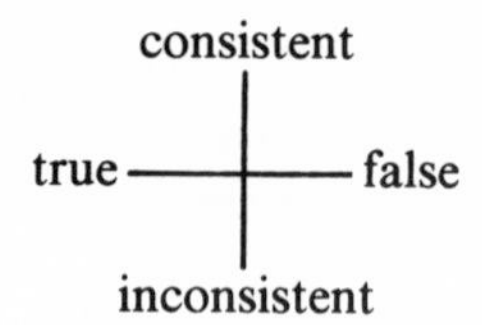

continues in this way until the witness makes some *inconsistent* statement. The lawyer can recognize the inconsistency and use it to smoke out the true, much as Solomon uncovered the true and false mothers.

In other words, true–false is not a true two-operator. There is another dimension that mediates it, or vice versa: true–false mediates inconsistency versus consistency also.

What, then, is the two-operator?

It is the operator which is not mediated by an independent duality. It differs from the dualities in fourfold analysis (quadratic analysis) in that it requires three dimensions to express it.

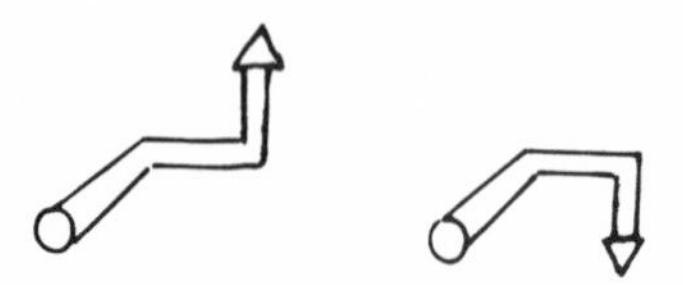

The two shapes above cannot be made to coincide. They have different chirality, right-handed and left-handed. This right- and left-handedness, like that of a right- or left-handed screw, is a basic property. One cannot be changed into the other by any manipulation.

Furthermore, this operator cannot be described objectively. It requires participation, which is projective.

We have already noted that besides corresponding to the directions of time, the chirality correlates to the two directions in which one can move in the cycle of action:

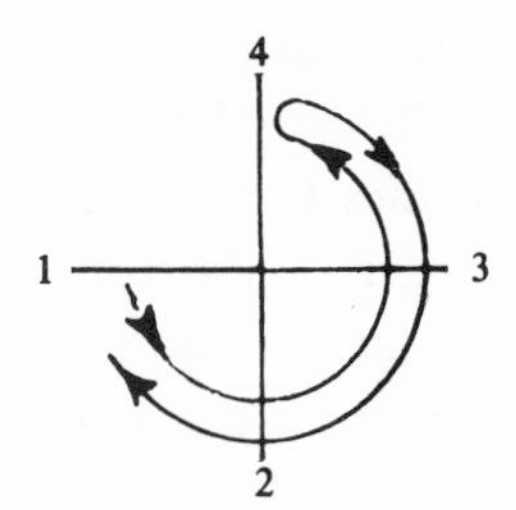

Here 1–2–3–4 (counterclockwise) is the natural direction taken, for example, in learning by trial and error; 4–3–2–1 (clockwise) is the direction of competence, of controlling a situation in which the laws of its operation are understood.

Consider a fisherman baiting a hook:

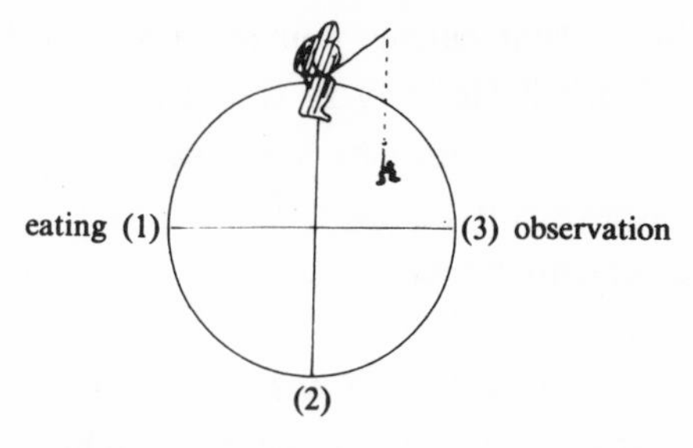

Let us assume the fisherman is very hungry and the bait is his last bit of food, let us say, a piece of bacon. Either he can eat the bacon, which would carry him to 1 in the counterclockwise direction, or he can use it to bait the hook, which would carry him to 3 in the clockwise direction, where he can watch the fish eat it. This, it will be recalled, is the "observation" point, which is opposite the "eating" point (1).

Then the fish bites and, in pulling back, hooks itself. It seems plausible to think of 2 as a point where the fish is caught, but the term *unconscious reaction,* which applies to the fish, does not seem appropriate for the fisherman, who reaches this point backed up by the whole learning cycle. The situation, however, is one of automation. The fisherman lets the situation work for him.

We already covered this turnabout in Chapter IX on purposive intelligence. The intelligence, having recognized that fire causes cooking, decides to build a fire in order to cook. Such an inversion of order requires the two-operator. This situation is so obvious that the effort to formalize it seems pedantic, and yet, if we don't, our science will continue to see the world in reverse.

For science sees cause producing effect because it *precedes* effect. But when we do something *in order to* obtain an effect, we make the effect the cause. We put the effect ahead of the cause. The prospect of the delicious cooked steak causes us to build the fire.

Aha! says the determinist, seeing a law. It is the stimulus of the cooked steak that causes you to make the fire! You are determined after all! He may in this manner save the letter of determinism, but determinism has now lost the meaning the determinist supposes it to have. It now yields to will and becomes the servant of will.

This is the most important point in the book, and it hinges on the two-operator, whose significance, in a human context, is the option to use the law to escape from the law. In the very midst of this thistle of determinism, we pluck the flower of freedom.

But the application of the two-operator is not limited to human situations. We have shown how, in the larger framework, it makes possible the beginning of life. It is the point at which the molecule starts to build order into itself and move against the flow of entropy.

While the two-operator is probably also the basis for the moral issue, that is an aspect best covered separately. We need first to get a thorough grasp of how it is used to master the law. Let us take a few examples from ordinary life. One I like occurred when the electrician had made a mistake and wired the two bells for the first-floor and second-floor apartments incorrectly. As I looked at the labels, thinking of the trouble of calling the electrician to change the wiring, my hand went to the labels. Hardly knowing what I was doing, I interchanged them. Then the wiring was correct.

Though the inventor not infrequently has to resort to some intuitional correction of this kind, it never becomes a formula, because to make the correction he has to invert the gestalt to see the whole situation inside out or upside down. My uncle, who had invented a sun motor which used mercury for its boiler, was told by the mechanic who was assembling the pipes, "The valve won't work; it floats (in the mercury) when it should sink." "Turn it upside down," said my uncle.

Or take the famous story of the days of the railroad tycoons. There was a price war between Gould and Vanderbilt in shipping cattle from Buffalo to New York. The rivals pushed the price lower and lower until Vanderbilt (let's say) had reduced the price to one dollar a head. Nothing further was heard from Gould, until Vanderbilt discovered Gould had bought the cattle and that Vanderbilt was shipping them for him way below cost. (Gould had reversed his role from shipper to owner.)

We here come into the moral question: is this taking advantage of the laws (as my friend called it) "kosher"? Here I must point out there is a difference between not taking advantage of the law because you don't know the law, and not doing so if you do. It is our first obligation to learn *how* to catch fish, how to make things work for us, and when we have done so, and only then, can we save our souls by not taking advantage of our fellow mortals.

XI The four elements

History assumes that thought has developed from a primitive state to the present advanced one.

This view may be justified in the case of science, but even here it is worth noticing that there is a difference between the body of science and the caliber of scientific thought. The body of science, depending as it does on the cumulative endeavors of all scientists, grows exponentially. The caliber of scientific thought, on the other hand, depends on persons, and rises or falls as individual talent varies. Even in persons, it may vary with age.

But something else goes on which a historical accounting would miss. Thought seems to become encumbered, rather than aided, by efforts to articulate itself. Someone asked the centipede which foot he started off with, and the centipede, trying to answer this question, became unable to walk at all.

In any case, we should study the disruption of innate understanding caused by the efforts of reason to justify its own legitimacy. In this sense, the growth of reason is a fall, first from the grace of instinctive motion, like that of a bird in flight, to that of a noisy machine, then to an erratic stagger, and finally, to a grinding halt.

The trouble began in Greece, with the "liberation" of thought from authoritarian teachings whose profundity rendered them too esoteric for "rational" explanation. These older teachings could never be explained, and such was their difficulty that the word "hermetic," which indicated their origin in the god Hermes, came to mean "closed off" or "sealed up."

Against this background of an unwritten tradition that was imparted only to initiates and understood by only a few, sometimes to be lost altogether, the use of reason must have burst like a flood of light. Certainly, the discovery of Pythagoras that musical notes bear the relation to one another of simple whole numbers was a triumph of rationality. ("Rational" literally means based on ratio, but it extends to mean any explanation which relates one thing to other things, especially to antecedents.)

But rationality is not always so effective. The ability of analysis to divide a thing into parts and compare them may engender absurdities when confronted with motion. Zeno's paradoxes—the arrow at every moment is at rest, therefore it cannot move; Achilles cannot catch the tortoise because as soon as he reaches the place where the tortoise was, the tortoise will have moved on—are, of course, absurd, but they are just as rational and make as much noise as better use of reason.

Indeed, the solution to the paradox of motion in the sense of a correct use of ratio had to wait for Newton. Meanwhile, there were brilliant philosophers like Berkeley to throw monkey wrenches into the machinery by insisting that a ratio between infinitesimals was an absurdity. We also had Hume at about this time, whose skepticism about the objectivity of causality we mentioned in the last chapter. That this should be considered significant by philosophers is an indication that the cogwheels of logic are highly vulnerable to monkey wrenches.

New impediments to thought are being discovered all the time. I will list only a few:

Dedekind's proof of continuity
Cantor's invention of a multitude of infinities
Russell's contribution to logic
Quantum theory's particle/wave enigma
Goedel's incompleteness theorem (an important step but still in its present state an impediment)

Still more recently, we have had other great discoveries. DNA, for example, while it undoubtedly does contribute to our knowledge, is responsible for an epidemic of nonsense subscribed to even by good scientists. We read that, any day now, science will produce Einsteins from a piece of his skin. (I hope later to show why this is a fallacy.)

Another is the computer and the madness it has engendered. Quite apart from that natural and perennial zeal for overdoing everything which characterizes American enterprise, computer mania is in the process of taking over people's imaginations. Science fiction looks forward to a future when its heroes will dispatch computers as their earlier counterparts did dragons. But what is alarming is that even good scientists (again) cannot distinguish computer activity from the mental processes of living creatures.

There are still other "remarkable advances" in science that, far from being an aid to thought, are incapacitating people for thought. Common sense has been so incapacitated by sophisticated obfuscation that we no longer even know how to pull our hand out of the fire when it's getting burned. In fact, it would almost seem that the progress of thought has reached a complete standstill, for we have not only more unanswerable questions and paradoxes than ever existed before, but a well-established staff of experts to preserve them for posterity and prevent any idle persons from presuming to solve them.

But it is for the outcasts, banned from the inner sanctum of science and without the prized qualifications, that I speak, for they have the necessary desperation required for the search I now urge them to make.

As Bacon said, it is the lighter stuff that floats on the river of time; the more profound sinks to the bottom.

Such has been the fate of a very ancient symbolic concept, that of the four elements, whose deeper meaning has long since disappeared beneath the surface. I would like to bring this meaning to those who, surfeited with the meaningless pyrotechnics of science, seek some basic guidelines to help find their way through the jungle of sophistry that is our present culture.

The four elements, fire, water, air, and earth, originated long before the Greeks, for when the Greeks referred to them, the manner in which they did so indicates that they did not comprehend their full meaning. The Greeks apparently thought of the elements as states of matter, just as most modern thinkers do:

Earth = solid
Air = gaseous
Water = liquid
Fire = igneous (ionized)

But this is not their *meaning.* To say air is gaseous and water is liquid is simply to reiterate the words themselves. We can expect more than that of ancient wisdom. We must look deeper. As often happens with symbols, we can actually find them in use right now. For example, the economist speaks of *liquid* assets, by which he means assets that are readily negotiated, like cash, as distinct from real estate, which is the frozen (or earth) form of investment. Anything concrete is earth, so the *element* earth means that aspect of a situation that is practical and "down to earth." Air, by contrast, is the mental aspect, and fire the vital and initiating factor.

These meanings emerge in the use of the words as verbs, thus:

To fire = to arouse, to start, "to throw out"
To air = to make known
To water = to nourish
To "earth" = to get down to earth; to materialize

This should give us a start, for it is not so much a question of defining the elements precisely, as it is of understanding the meanings in terms of their classically assigned quadrate relation, with air opposite fire and water opposite earth.

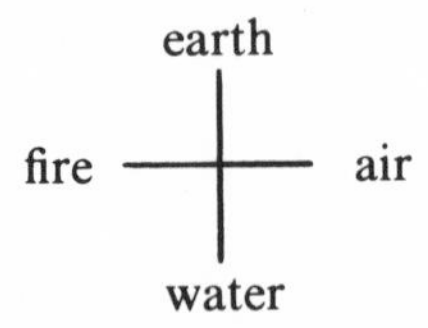

The elements designate directions in space, as they do in the signs of the zodiac. These directions in space are also associated with people's poetic understanding of the seasons. Thus, taking the four signs of the zodiac assigned to the equinoctial points (the first day of spring and of fall) and the solstices (the first day of summer and of winter), we have four signs: Aries or spring, Cancer or summer, Libra or fall, and Capricorn or winter.

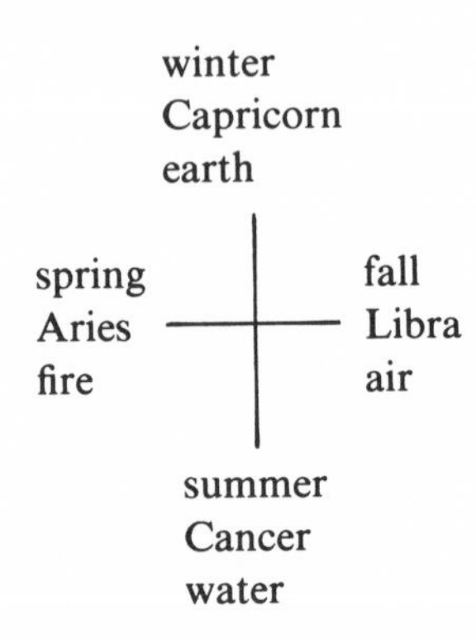

Spring begins the yearly cycle, just as the east initiates the diurnal cycle; thus people associate spring with the east. The astrological sign for spring is Aries (fire). It signifies increasing heat. Its meaning is spontaneous outrush, symbolized by the ram which charges with its head down.

The outrush of Aries awakens the response of summer growth represented by Cancer. The floodwaters of spring assume their nurturing level. Cancer is symbolized by the crab with its protective shell for the growing form within.

The autumnal sun is in the sign of Libra (air). The harvest is weighed, evaluated. Autumnal winds take the turning leaves. It is the time of mental awakening and judgment, symbolized by the scales.

The sun enters Capricorn (earth) in winter, when its heat is at a minimum. Dormant nature gives us the meaning of conservation and control. Lessons learned in the seasonal cycle are absorbed during the winter; the watery element is congealed. Capricorn's symbol, the goat, denotes the surefootedness of this animal, but other goat-like references are available: the scapegoat that voluntarily takes the "blame," the self-assurance of the goat, its ability to digest, and so on.

Spring and fall are opposites, as are fire and air; so too summer and winter and their associated elements. Our temporal awareness of the seasons finds correspondence in the spatial notation of astrology.

These astrological meanings, it will be noted, fit the four types of action in the learning cycle at the end of Chapter II. On the left, corresponding to the first day of spring, is *acceleration or unconscious action,* precisely what is happening in spring. At the bottom we showed velocity, which is change of position, but it will be recalled that we generalized this to cover all *unconscious reaction.* On the right we had position, which we equated to observation, *conscious reaction.* This is the function of Libra, looking to see where one is at. At the top we had control, or *conscious action,* which is as good a meaning for Capricorn as can be made in two words.

Quite apart from the validity of astrology, we can therefore see that the signs denote, and derive their meaning from, the critical points in a cycle dealing with velocity and acceleration. Relating this to our subject, the geometry of meaning, we can see that there is an underlying and abstract fourfoldness in the signs.

Note that the four meanings are defined by one another. Unconscious action on the left is doubly opposite conscious reaction on the right. Likewise, the two cross products, unconscious reaction and conscious action, are on bottom and top.

This should make it clear that the elements occur together. Like the directions in space, they coexist. One can emphasize one element at the expense of another, just as one can travel west instead of east, but this does not mean the other element or direction ceases to exist. One cannot imagine space extending only to the west. Space is *omni*directional, and so is everything else. It is true that a thing moves only one way at a time, but the cyclic nature of all occurrences makes it inevitable that the complete development of anything takes it through all four phases. Like the pendulum we started with, the phases of the cycle derive their nature from this sequential interrelation.

The elements, as expressed in the cardinal signs which we have just outlined, are thus four kinds of action. For example, the action of the soldier typifies Aries (Mars), or fire, as compared with the observational action of the scientist, which typifies Libra, or air. The engineer or manager, who makes and controls, contrasts with the passive consumer, much as Capricorn (earth) contrasts with Cancer, water.

But these are all forms of action. To fully explore the elements, we should draw on the other two sets of four, making three signs in each element.

Taking them in sets as we did in the case of the measure formulae, the next set are what are known as the fixed signs, which "incorporate" the activity of the cardinal signs and produce *states:*

	Action (leads to)	*States*
Fire	Outrush (Aries)	Being (Leo)
Water	Change (Cancer)	Transformation (Scorpio)
Air	Observation (Libra)	Significance (Aquarius)
Earth	Control (Capricorn)	Establishment (Taurus)

We can also place states in opposition, as we placed measure formulae in opposition. Transformation as the opposite of Establishment gives us no difficulty. But Significance as the opposite of Being may seem strange.

Here is where the deeper insights afforded by the elements are a help, for in recognizing Leo (being) as fire, we may characterize it as centrifugal, *throwing out.* This, in fact, is what the basic symbolism suggests, for the sun is symbolized by a circle with rays going out from it. The lion has probably become its representative because his mane, springing from his head, also depicts an outgoing, or centrifugal, motion.

The opposite tendency or mode would be centripetal, converging to a point. The astrological symbol of Aquarius is the water bearer, sometimes shown pouring water from a jug. That Aquarius is an air sign, not a water sign, makes sense if we realize that Aquarius represents precipitation or rainfall—the return to center of that which heat has caused to evaporate—and hence the opposite of Leo.

Another probable reference to Aquarius suggested by Eric Schroeder,* an authority on symbols, is the unicorn, a mythical beast whose single horn twists to a point, and thus literally portrays the centripetal. Its opposition to Leo is shown

**Muhammad's People.* Portland, Maine: Bond Wheelwright Co., 1955.

in the British royal coat of arms. As the nursery rhyme says, "The lion and the unicorn are fighting for the crown." This convergence to a point, or *focus,* of the power of the head (the horn) confirms the mental nature of this air sign and stresses its meaning as significance.

Since the contrast between centrifugal and centripetal forces is clearer than that between words like "being" and "significance," we are justified in using it to explain or define their meaning. The contrast between centrifugal and centripetal implies oppositeness, and hence exemplifies the geometry implicit in meaning.

We can discover a similar oppositeness between establishment (Taurus, earth) and transformation (Scorpio, water). Here the point is to recognize these states as possessive (in contrast to the states of being and not being). Establishment is having, and transformation its opposite, not having. In a negative sense, this means the destruction of the former, but in a positive sense, its reconstitution. Taurus, or the bull, is the zodiacal symbol and corresponds to the month of May. We can sense the "bull market" feeling of this time of the year when the spirit of increase is bursting forth everywhere. The opposite sign, the scorpion with its poisonous sting, does not so much symbolize destruction as the intensity of reappraisal necessary to transformation. Death, which is associated with this sign, is destruction, but only of the form, since there is in cosmic symbolism no true death. This vital renewal is emphasized in the other symbol of Scorpio, the eagle.

There remain what are called the mutable signs, which, like the third set of measure formulae, have to do with kinds of relationship.

These are Gemini (air), Virgo (earth), Sagittarius (fire), and Pisces (water). All except Virgo deal with forms of two-ness, and this recognition makes it possible to relate them in quadrate relationship. Thus Gemini, or the twins, are two out of one (mother), whereas Sagittarius, the archer astride a horse shooting an arrow, is one (arrow) out of the two. Since Sagittarius rules the hips, the point where the two legs come together to form the spine, this again is one growing out of two, and hence synthesis. Similarly, Virgo is one out of one, and Pisces, the two fishes pulling against one another, are two out of two (which helps bring out the tendency of this sign to confusion and delusion).

The word pairing which we used earlier in the book for the four forms of relation is also apt:

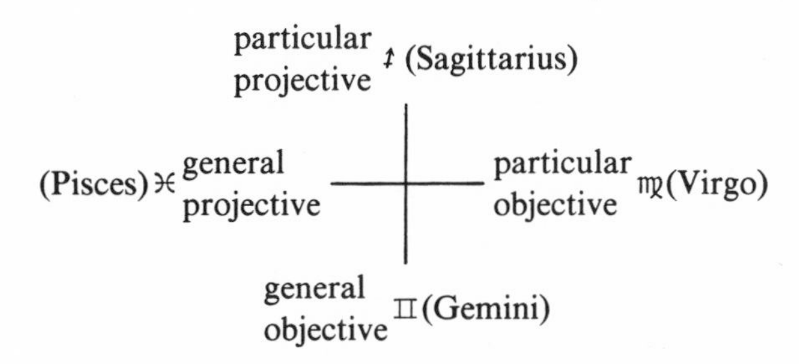

Sagittarius, the arrow, is the *particular projective;* Gemini opposite it, the *general objective.* The criterion for objectivity in science is that an experiment be repeatable. One must be able to "duplicate" it and obtain the same result, an emphasis suggested by the twins.

The *particular objective,* which we earlier related to empirical fact and to the work required to determine it, corresponds to Virgo, which is the sign that stands for discrimination, separation, purification, and in astrological usage rules the "house" of work. (The "houses" are the twelve kinds of relationship possible to the self, and have a correspondence to the signs.) "Work," it will be recalled, is the equivalent in science to energy, whose measure formula is

ML^2/T^2, and was positioned at four o'clock, which is where Virgo falls in the conventional disposition of the signs.

Pisces, as opposite Virgo, is lack of discrimination, but its positive reading is *faith.* There is also in Pisces the connotation of sympathy, of going along with, which is again a positive reading of the opposite of the separation and purification (negative hypochondria) implied in Virgo.

We can now survey the three sets or modalities—cardinal, fixed, and mutable—which combine with the four signs or elements, fire, air, water, and earth, to produce the twelve modes of being set forth in the zodiac.

The origins of the zodiac go further back in time than Greek philosophy, or even the I Ching. The zodiac is, of course, the basis for astrology, currently regarded as a pseudo-science (largely because its critics misinterpret what it actually purports to do). Apart from the question of whether astrology is or is not valid, one can hardly question the zodiac. Spring *is* the time of physical acceleration; autumn is the opposite, the time of mental stimulus. Summer is the time of physical change, and winter the time when growth has ceased.

"So what," says the modern mind, putting its faith in the constructions of science.

But, as I have been at pains to show and have obliged my patient reader to witness, the measure formulae, which are the very basis of science, draw their meaning from the same angular relationships, from dispositions about a circle, from the opposition and complementarity that thousands of years ago were evident to the ancients. This circle of the zodiac (*not* constellations, which are quite arbitrary pictorial projections hardly even resembling the stellar configurations to which they refer) is *basic.* It is at once the cycle of action, the progression of the seasons, directions in space, the measure formulae. Even more importantly, it is the *whole* whose various aspects are the signs, the measure formulae, or the seasons.

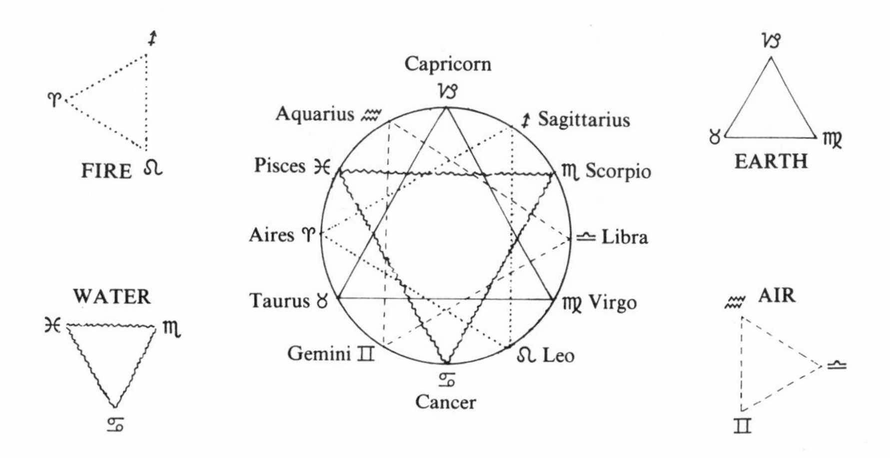

And because it is the whole (and does not let us forget it), it is a more valid reference than science, for science, which is knowledge, not wisdom, is not the whole story. Scientific knowledge, in fact, can be rather mischievous and even dangerous in that it is at best partial. In the next chapter I propose to show how, with the help of the four elements, we can break the spell which the evil magician has cast over the enchanted princess.

XII Free will

I have recently been reading Ernst Cassirer's *Determinism and Indeterminism in Modern Physics.** Cassirer is a brilliant writer. He is thoroughly at home in science and in philosophy. He is an exponent of the new enlightenment and, generally speaking, provides as fair an example as I could pick of the obfuscation and ambivalence in which modern thought becomes entangled.

The book leads up to a final chapter in which Cassirer deals with the crucial question of whether the indeterminism established by quantum physics has a bearing on the question of free will and the related questions of ethics. He concludes, as have almost all authorities who have approached the problem, that it does not.

Of course, I believe it does, but because I am outnumbered and outpointed by rational argument, for my opponents are all rational men, I am compelled either to retire in defeat or to meet their arguments eyeball to eyeball. This is difficult, if not impossible, for the use of reason, which has trapped them, would also trap me. There is nothing I'd enjoy better than a blow-by-blow *rational* argument, but even if I were to win the battles, I could not win the war, as I've learned many times.

*New Haven: Yale University Press, 1956.

Therefore, instead of trying to win the argument in regard to free will, I propose to lay siege to philosophy in general, in order to restore that whole thinking with which we have lost touch in our bedazzlement with science.

We can begin with the famous Laplacean "intelligence," which, given accurate information about every particle in the universe, could predict what was going to happen for all time. Cassirer devotes his first chapter to this subject.

He explains that his reference to the Laplacean spirit is not made because it is appropriate as a picture of a universe, but because it is not. For were it possible for human understanding to raise itself to this ideal intelligence, it would still be but "one aspect of the total of being." Here we can back up Cassirer, for of the three modalities—relation, action, and state—all the knowledge of the Laplacean spirit is confined to relationship, and of the four kinds of relationship, it accounts only for the self-contained, or objective general, type of information. So we can say that even if such a degree of knowledge were possible, it would have access only to one-twelfth of reality.

So far, so good. But Cassirer now says that the vast and important domains contained by reality dissolve into nothingness when we depart from this idealized objectivity. Perhaps I misunderstand him, but I would insist we do have access to these vast and important domains, not only through the other kinds of knowing, emotional, intuitive, etc., but through four kinds of action and four kinds of experience of states. In fact, the notion of objective structure can never do more than share totality with eleven other aspects.

But perhaps I should formulate better arguments about this important issue, for it emerges again and again, and in many forms, in Western thought. Cassirer himself, of course, realizes the shortcomings of the Laplacean intelligence. He shows that Hume was able to shatter the monolithic unity of the system of cause and effect (upon which the determinism of the Laplacean intelligence depends) by simply doubting the objectivity of causality. Hume's skepticism was based on the fact that we know about the objective world only through our perceptions, which is where the principle of causation exists. We can find no evidence of causality "out there." *It is a mere belief.*

I find Hume's argument unconvincing. While it is true that cause and effect is not objective because it is a relationship that requires time, and hence not the symmetrical type of relationship that objectivity demands, it is not for this reason less "real." In fact, according to the ontological scheme we have set forth, cause and effect is second-level. It is of the same ontological order as is time. It is also at the level we accord to emotion, things felt, which are the "realest" of all, for they include pain and pleasure.*

So where do we stand? The "shocking" skepticism of Hume, which affirms that cause and effect is not objective, is already anticipated. And the supposed implication, that the reality of cause and effect is therefore undermined, is wrong-end-to. One just doesn't undermine reality. One discovers a deeper reality, both more real in feeling and ontologically prior because it is second-level.

This is not Cassirer's view, however. He calls the inability to give up the "belief" in causality a mental failing. He feels we "overstep the limits which have been set for human knowledge as soon as we give this belief any objective basis."**

We cannot blame Cassirer for the confusion that Hume initiated; it is of long standing. It consists in proving that because x is not objective, it does not exist. If x were a

*The reader will note that I am not trying to do justice to the arguments of our guest philosophers. I am, rather, collecting these arguments and assigning them to a prepared system which already anticipates their message and implements it.

***Ibid.*, p. 16.

headache, its "nonobjectivity" would diminish its reality not a whit. Why then should the nonobjectivity of causality diminish its reality? We are being given the same razzmatazz that made Zeno famous.

But we can see the situation as one which calls for application of the three-operator, not analysis. Cause and effect involves time sequence. Referring to Chapter XI, the relation of fire to cooking is causal if fire comes first. This is the natural order when the barn burns down and roast pig results. But purposive intelligence can invert the relation and build a fire *in order to* cook. It is entirely correct that causation is not objective. If it were, the order could not be inverted, as it must be if we build a fire in order to cook.

Cassirer now introduces Kant. He says that Kant confirms Hume, but takes the argument still further in that he applies it to the causal concept *in general.* Kant concludes that we cannot through reason conceive how *any* change comes about. Addressing himself to the epistemological question, Kant entitles *transcendental* all knowledge which is occupied, not with objects, but with how knowledge of objects comes about (my paraphrase). It is *a priori* to the objective knowledge.

This term "transcendental" has always puzzled me, partly because the word seems to imply something "beyond what is natural." Seen in the light of our present discussion, however, the word is equivalent to what we call projective, that is, it is nonobjective; it is prior to objectivity. I see no reason for not thinking it natural. It is the *basis* for (but not equivalent to) rational knowledge.

Perhaps the simplest way to see the difference is to correlate projective (Kant's transcendental) to *quantity,* and objective to *ratio.* The difference is basic and all important: the former we have correlated to material cause (water), and the latter to formal cause (air). It is also the old issue of substance versus form. It gave rise to Aristotle's question: is the soul substance or is it form? His decision, reached in his later work, was that it is form. If so, it follows that since form can have no existence apart from a body, the soul does not survive the body. It is not immortal. Since the earlier doctrine that the soul is substance, and therefore immortal,* was not invented by Plato, but was traditional, we can say that this reasoned conclusion of Aristotle initiated the emphasis on objectivity which has characterized Western thought.

For form *is* communicable, it can be defined, it can be formulated. Moreover, because it can be formulated, it can be conceptualized. If we ask reason to tell us what is real, reason will invariably present us with its own produce. It will hand us concepts. Substance it will deny. This takes us to another part of the forest, for during the period that Hume and Kant were demolishing the objectivity of cause and effect, other philosophers were busy demolishing substance. As we have already pointed out in Chapter VIII on substance and form, Berkeley proposed to dismiss the notion of substance altogether. He pointed out that the chemist has no need for the notion of substance since he can always tell whether the object before him is gold by comparing it in various ways with other objects. He obtains its specific gravity, its combining proportions, its solubility in different acids—ratios, in short.

But the good bishop was playing two chess games at once. He must have been quite a cantankerous fellow, for he was not content to elevate ratio at the expense of substance, as Aristotle had elevated form at its expense. He also saw fit to attack Newton for doing the same thing but more competently and in a more appropriate application.

We have already touched on this subject in Chapter II, where we briefly referred to Newton's discovery that not only could we measure space and time, we could employ the ratio

*An early hint of the conservation of mass–energy.

of space to time, that is, *velocity.* But Berkeley insisted that such a ratio was a "logical absurdity." It is true that it was perhaps the device of taking this ratio over infinitesimal intervals of time that disturbed Berkeley, yet this was the step that freed the ratio from the necessity of a tie to substance, the very thing Berkeley elsewhere insisted on.

Now isn't that strange? Berkeley, reasoning shrewdly, decides substance is superfluous, as it is for chemists; yet when Newton takes the step that eliminates it and formally elevates the Bishop's own proposal by the invention of the calculus, the Bishop is outraged. He would deny substance to chemists and ratio to mathematicians. This kind of "push me, pull me" is great stuff to a higher form of consciousness, but it is murder for philosophy because the philosopher feels he must take sides.

But bear with me. Johann Sebastian Bach, who was born March 21, 1685, the day before Berkeley, at a time when a number of planets were in sharp opposition, translated this opposition into his contrapuntal music. He was able to stand above the "push me, pull me," and sense it as an ecstasy, a cosmic copulation. Berkeley was victimized by it.

But we must not tarry with astrology. My purpose is to show the interplay of the four elements, fire, water, air, and earth, the necessity of opposites for one another, and of both for that which mediates. In passing, we have also seen how these great issues of philosophy are built into the fabric of reality, which is not form or substance alone, but both together, and more, for we must not forget their quadrate relationship with function and the physical object.

Put another way, we can say that in our materialist orientation (which, by the way, I am not criticizing, because one can start at any point in the circle or cycle) we begin, or at least agree to begin, with *physical objects or particulars of experience.* We cannot rest there because we need to

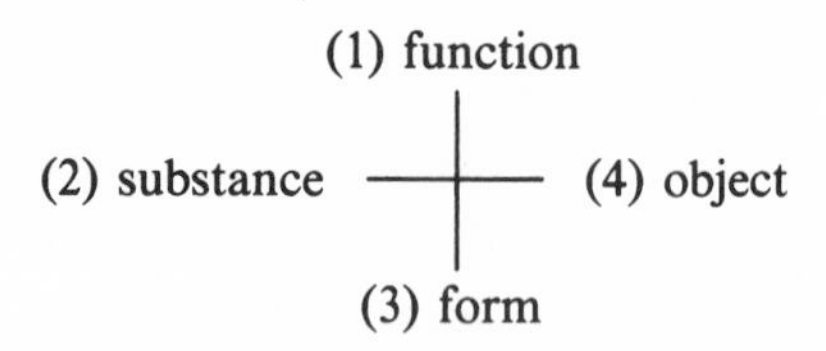

extrapolate, to theorize from what we know of these particulars, to something else. We have two alternatives. We can extract from the objects their *form,* which includes formulating their motions, behavior, nature, etc., or we can extract from their substance, their *feel.* The former leads to a static view of relationship structure, a world of interlocking connectivity which can be seen all at once, like a map of the subways in a great city, a map that *includes* time, but does so in a way that removes from time its excruciating reality as change. This is the mental world, the world seen by the Laplacean intelligence.

The latter, or the emphasis on substance, leads us back to the feeling world that precedes intellect or formulation. Lacking the light that reveals things all at once, it might be thought that it is a world of darkness, but it is not. It is full of images, spooks, wraiths, phantoms but also of beauty; it is the world of dreams, of images. It is, in a sense, the world of illusion, but it is the movement of life because it *is* substance, we can take hold of it.

The deficiency of this world is that it is both a Garden of Eden and a trap. To escape, we must sever ourselves from it by an act of the mind, like a fish climbing out of the water to get its bearings, and from this island of the ego, we can look about us and learn a little.

It is impossible to have either form or substance in an absolutely pure state. Except, perhaps, for raw energy itself, there is no substance that does not have some structure. The raw material of which we make things, wood, iron, clay, has structure in its molecular interstices, and even the purest mathematical forms require substance for their representation; we need pencil and paper, or electricity to run the computer.

So the endless debates of philosophers on form versus substance can never be settled on the terms that the contestants desire: a unilateral decision for one and against the other.

If the reader understands what we have covered so far, he may close the book. But first let him be prepared to use the tools we have resurrected from the ancients. For we are not living in the 17th and 18th centuries, when the Age of Reason wore a peruke and graced its halls with colored silk and string quartets. Nor even in the 19th century, when prisons and policemen were invented, when steam locomotives, stoves, and top hats led the somber parade toward morality and transcendental philosophy. (A deadening era too, since the testimony from science that broke the rule of reason was the second law of thermodynamics, or entropy, stating that the universe is running down, down, down, to end forever in a lukewarm death.)

We are living in our own time. We have made new discoveries, found many things not known before. It is already evident that these advances in technology are a mixed blessing, for reasons that are hard to generalize. But I think it is safe to say that their bad effect proceeds from the fact that they are partial. They elevate and amplify one factor at the expense of others. Not long ago, writers were given to saying that man lagged behind the developments of science, implying that some more advanced type of humanity was necessary to these new devices, but this notion, which puts the cart before the horse, is less in evidence of late. We are aware that pollution of the atmosphere and hydrosphere is a fault of the overall control of technology, not of man's evolution.

In any case, this is not our present concern. We are

concerned with mental pollution, with the degree to which new ideas have so deluded us that we have lost all common sense. I am not against science. I am all for it. Nor do I believe in ignoring recent progress in science, for the handwriting of quantum physics contains an insight more profound than any it has been man's privilege to hear in the many centuries of modern history. But because man's mind is so deluded with rational half-truths, we have not read this message correctly.

How may we do so? Let me return to Cassirer.

No one would question his merits as a mind and as a philosopher. And as I read *Determinism and Indeterminism in Modern Physics,* like watching an archaeologist digging up a valuable find, I saw how carefully and rightfully he made his progress, only, at the very end, to miss the point.

But, as I said earlier, he is not alone in this. If I were to say, "Hold on there, Professor. That lump of clay which you just tossed aside is worth more than your whole museum. Please look at it closely, scrape off the encrusted matter, test its hardness on your watch crystal. It is the jewel you seek, the diamond that can cut the hardest steel, that will cleave carborundum like butter," there are dozens who would say I was wrong. Therefore, in this chapter I have taken the reader on a long excursion. Not being able to climb the north face of the Matterhorn, I have taken him all the way around. In this whole book I have laid down a scheme of meaning which, in a sense, is just a map upon which to base the strategy for the climb.

It is here that the challenge for the reader lies: to cut through the obscurity of the present, not the smog of pollution, but the smog of education, of intellect that has lost touch with both sense and intuition. How are we to restore man to wholeness?

I was told of a Warhol movie, *Trash,* in which an impotent junkie is the protagonist. A number of females endeavor to induce an erection, using techniques currently described as oral–genital. Of course, I am shocked, etc., but on reflection, I recognize the outlines of a very ancient myth showing through, the myth of Osiris, the man-god who descends into the lower world, where he is cut up into pieces by Set. Set throws the pieces into the marsh. Isis, Osiris' sister and mate, gathers them up and finds all but the penis. Despite this deficiency, Isis conceives from the assembled corpse the infant Horus, the hero who conquers Set and becomes the Sun God.

What is the penis of Osiris? We have by this time become accustomed to the Freudian ploy: "What is the meaning of (the Tower of Pisa, climbing Mount Everest, a golf club, kitchen knives)?" "A sex symbol," says the shrink. But now we have a myth about the penis. What does that mean? We can no longer say simply that it is a sex symbol. That tells us nothing. Taking the myth in its totality, we can see that what is lost when Osiris is dismembered is wholeness. Like a car disassembled and no longer a car, Osiris is not capable of generation, not because he lacks a penis, but because, since he is not a whole person, he is not able to initiate, to *be cause.* This, then, is the meaning of the penis. It is the power of generation, and it is the person in his capacity to initiate new endeavors, to create, to make things happen.

The myth of Osiris is a death-and-rebirth myth. It is a hero myth, but it is also a virgin birth myth, which is to say, it tells how something comes out of nothing. It is about *first cause.* In several ways we are told that the cause of generation is not a visible physical thing, not something objective.

Quantum physics, I believe, tells the same story. It tells us that behind the phenomenal objective world there is something we cannot in any objective sense know, something which can be characterized only as *uncertainty,* but something which is the cause. It is a *quantum of action,* a whole act. Cassirer, as I have said, rejects the possibility that the indeterminacy discovered by quantum physics is the freedom of free will. Because I do not agree, it is hard for me to state his reasons fairly, but I will try.

Reaching the end of his exposition of causality and of quantum theory, Cassirer takes up the free-will question in his "Concluding Remarks and Implications for Ethics."* He begins with the stock "hands off" warning to which professional scientists resort when they want to keep philosophers and amateurs from prodding into the significance of concepts which they themselves use in a rather specialized manner. Kant, he says, insisted that only distortion occurs when we permit boundaries between fields of knowledge to run together. This is what occurs, says Cassirer, when statements about indeterminism in quantum theory are directly connected with metaphysical speculations about freedom of the will.

He then goes on to say that it is precisely the uniqueness of ethics that sets it in a realm apart from that of physics. Ethics, he says, would be in a bad way, would lose all its dignity, "if it could maintain its authority and fulfill its particular function in no other way than by keeping a lookout for gaps in the scientific explanation of nature and taking shelter, so to speak, in these gaps."

The eloquence of this point is admirable, but in considering such basic issues as that of first cause, we must not make divisions; we cannot truly separate the world into a physical and an ethical domain without having made an irreversible choice in the limitations of the respective fields. Moreover, Cassirer is mixing up the *methodology* of physics with the *findings* of physics.

It was not a part of the *method* of physics to conjure up indeterminacy. Indeterminacy was a shattering discovery that delivered a knockout blow to classical science by showing that

**Ibid.*, pp. 197–198.

the method of science could go only so far. There was vital core in phenomena, a core that could not be formulated or bounded. This finding is of concern to philosophers because it indicates that the boundary separating ethics from physics does not exist at the primordial level. The revolutionary discoveries of quantum physics are a reminder that even the "particles" of classical physics are engaging in ethical struggles. Having confused the findings of science (which are as much revelations as are those of religion) with the method of science (which is a code of behavior, a sort of ethics), Cassirer now goes on to say that what he calls the "negative concept of indeterminacy" is inadequate for the positive nature of moral freedom.

By now we should be on guard about positive and negative. We recall that every attribute has its positive and negative form, and also that each attribute (in the circle of meaning) is the opposite of the one 180 degrees from it. Thus faith and discrimination are opposite and, in a sense, the negation of one another.

To apply this concept to indeterminacy, let me tell a story. During experiments with a very remarkable "sensitive," Frederick Marion, whose special skill was psychometry, or the ability to tell the meaning of a word or message written on a piece of paper and folded so that the subject cannot see it, I asked him to use his talent to tell me the meaning of symbols of which I myself did not know the meaning, among them astrological symbols. I gave him, folded, the sign of Sagittarius. Marion concentrated for an unusually long time. Finally, striking his fist against the palm of his hand, he said, "This is the power of ignorance!" My mind went to the emblem used on the Sikorsky helicopter:

> "The bumblebee, according to the theory of aerodynamics, cannot fly, but since the bumblebee knows nothing about aerodynamics, it goes ahead and flies anyway."

Do I make myself clear? On the circle of the zodiac, Sagittarius is opposite Gemini. Gemini is knowledge. Sagittarius is nonknowledge, ignorance, but it has a positive connotation, that of taking the plunge, making a decision, taking off, like an arrow launched into the unknown. To top it off, we have the position of Sagittarius (one o'clock) as that of *action,* the measure formula for the quantum of action (see chart at the end of Chapter IV).

In other words, the concept of indeterminacy, negative as it may be to the conscious mind, is positive in its own right. We have, in fact, already equated mind to negative being, so it follows that nonknowledge, or ignorance (uncertainty, indeterminacy), is positive being.

But there is another issue involved. Cassirer seems to equate free will with moral freedom. Here I think, thorough as he usually is, he has skipped a step. What is essential, especially in view of the deterministic bias which has infected the age, is to establish the possibility of free decision. We admit man's many failings, his animal needs, his vanities, his shortcomings, but is there or is there not somewhere at the core of his being some essential capacity to *be cause,* to initiate a course of action that is not "fathered" by any prior cause?

That there is such a capability is testified to by all the great religions. This is the true meaning of the virgin birth. Osiris is reborn as Horus, who conquers Set. Christ is born of the Virgin not because of some biological freak, but because the Son of God can be fathered only by the ineffable. We are all, in this sense, the sons of God, the ineffable, the nonobjective. (If we have a spiritual rebirth, it is not due to some exterior cause. It has no father, no authority who tells it to be.) It is the same whether we call this capability first cause or whether we call it divine cause. I realize the resistance most readers will throw up against this notion, and this resistance is paradoxical: the very refusal to accept ulterior reason is an affirmation of self-determination, leading to self-realization and self-transformation.

This is my own wording for an ineffable mystery. It is perhaps what Cassirer means by moral freedom. But, as I said, I think a step has been left out. First, the new self must be born; the moral step comes later. It is perhaps described in Christ's temptation. I am not a Biblical scholar, but from a number of other considerations, it would appear that free will must first learn to manipulate matter (recall Chapter IX on purposive intelligence), and when this task is done, when knowledge is gained from the earth experience, the moral issue emerges.

I am concerned here with free will. The additional factor involved in moral action is beyond our present inquiry—and certainly beyond my abilities—but I would surmise that moral action is just about as indeterminate as is the quantum of action. The example given by Cassirer from the *Phaedo,* in which Socrates chooses to take his punishment, is certainly a moral action; but not because it is determined, either because the Athenians decided on punishment, or because Socrates complied with their decision. Some moral acts are in conformity to the state and some are not, and there is no *predetermined* answer.

Cassirer seems to imply that an intelligent man acts in a predetermined way, but even if this is so, it is not the kind of predetermination dictated by the laws of matter. Before this moral act, with its own predetermination, if such there be, there must be the possibility of a completely free decision.

This would certainly require complete indeterminacy, and the fact that indeterminacy is found by quantum physics at the core of what was once called matter is significant both inside and outside the boundaries of physics. It is a discovery about the world, our world, for we all exist in one world. The atoms studied by Mr. Bohr are the same atoms that compose my body. If they cannot be predicted, there is that same free element in me. We can have no boundaries here.

The question, not raised by Cassirer, but cited by others (notably Caws* and Waddington**), that the quantum of action involves much too small a quantity of energy to be adequate for moving my fingers and thumbs, is easily answered. The quantum of action, in all life process, is a *trigger energy.* It does not need to have more than a certain minimal value (enough to change a molecular bond) to control the living organism. It does so through an elaborate hierarchy: molecules control cells, cells control nerves, nerves control muscles, muscles control bodies, and (we can go on to say) bodies control bulldozers and B-52's. But I will not expand on this here. It is a straightforward mechanical problem.

Others reject indeterminacy as a basis for free will on the grounds that if we allow indeterminacy, anything could happen, nothing could be relied on. This is doubly mistaken, since in the first place, there is no option about accepting indeterminacy. It is a fact of life. In the second place, the indeterminacy of quantum physics (which exists) does not, in fact, create a general breakdown of laws. The indeterminacy of quantum physics is that of *individual* electrons, atoms, and molecules. It does not affect the laws of aggregates. The vase stands still despite the jostling of its individual molecules. Only for life, hierarchically organized, does this "individuality" have wider scope and become able to cause macroscopic uncertainty (negative entropy for plants, and mobility for animals).

The objection that nothing could be relied on if indeterminacy prevailed applies more properly to man-made mechanisms in which provision against the failure of a part can be obtained only by an adequate factor of safety. Even

*Caws, Peter. *The Philosophy of Science.* Princeton: Van Nostrand, 1965.

**Waddington, C. H. *The Nature of Life.* London: Allen & Unwin, 1961.

with this insurance, few machines have an infinite life. But these are mechanical questions.

Our concern is not with mechanical problems, but with the much more difficult intersection of the spiritual and structural worlds. Cassirer now takes up Spinoza. (Yesterday I saw a bulldozer labeled SPINOSA.) Cassirer seems also to express the notion of one world we mentioned above, for he says that Spinoza's method demands that we do not treat human actions as a "state within a state." "There is only one order and one law underlying happening, as truly there exists only one being, one all-embracing substance."*

Here it might seem we've won our point that the indeterminacy of physics is significant for free will because both must be in one and the same world. But Cassirer surprises us. His very next sentence is: "The concept of purpose must therefore be excluded not only from science but also from ethical considerations." I would have included it in both! It is obvious that Cassirer does not think of purpose as equivalent to indeterminacy. This was our basic starting point (the kind of knowing described in Chapter I as purpose or final cause).

Cassirer now is really in a fix. Having banished purpose from nature and from man, he has to open the door a crack and let it back. He says, "Even Spinoza cannot evade the admission that there is a 'different kind of cause' to which we are led when we are considering human actions." The footwork now becomes very rapid. He continues, "We cannot nor do we wish to withdraw from the dominion of general laws of nature; but these take on a different character when they refer not to the motion of bodies, but to our self-conscious activity . . ."**

*Ibid., p. 200.

**Ibid., p. 200.

So now we have two worlds again, seen from within and seen from without. This is precisely the point; purpose as seen from without is indeterminacy. What I can cause to happen by my free will, the observer can know only as indeterminacy. If, on the other hand, I cannot cause anything to happen, the observer says I am determined. In fact, *I am dead.*

But back to Cassirer: "Ethical laws are natural laws, but they are laws of our rational nature . . . To act freely does not mean to act arbitrarily or without prior decision; it means, rather, to act in accordance with a decision which is in harmony with the essence of our reason."*

And now a word from our sponsor. (I'm sorry, but I simply cannot let that pass.) Here Cassirer makes a double oversight. In the first place, he confuses the initial decision with the organized action that must follow if the decision is to be effective. As we pointed out earlier, freedom of action requires freedom of decision *plus* a whole lot of determinate agencies to carry it out. Second, the phrase "in harmony with the essence of our reason" contains, I am afraid, a most serious error. For if reason is to dictate "free" action, it is not the action we are talking about, the action of first cause, which is the creative or novel type or action that cannot be anticipated by reason. (Just as "novelty" is defined by the patent office as that which could not have been anticipated by one skilled in the art.)

If, on the other hand, we take reason here to mean a higher type of reason, something to which only spiritual intuition has access, that is something quite different from the logical connectedness of ordinary reason.

And it is evident that this is what Cassirer refers to, for he goes on to say that the "concern [of reason] is to understand the whole . . . its realm is not that of mere existence but of pure essence . . . it is the *amor dei intellectualis* . . . He who is

**Ibid.*, p. 200.

filled with love for and insight into the whole does not succumb to the illusions of the imagination or the incitements of passing and momentary motives." And, we would add, does not succumb to *rationalization.* In short, this is not reason at all. This is the medieval *intellectus,* not the modern intellect (a 180 degree reversal in terms of our geometry). It is the higher mind, the intuition.

What interests me especially in the above is that it ties in with yet another modern variant of the Zeno-type "paradoxes" that encumber free will, the notion that moral law and physical law are somehow equivalent. To name names, I will add that it is the view taken by three scientists in recent works: Waddington's *Ethical Animal,* Hardy's *The Sacred Flame,* and Margenau's *Science and Ethics.* What is common to all three authors is that as men of science they have come to believe in the laws of nature which, having their own beauty, produce the order of the phenomenal world, and these authors view moral behavior as similar. For Hardy, it is social laws that guide men; for Margenau, it is the determinism or physics (Margenau wrote the introduction to Cassirer).

But, as we said of Cassirer, this admiration of natural order confuses the ends and the means. Free choice, if truly free, is indeterminate, but the means that the will uses to carry out its objective must be predictable, or things won't turn out as planned. My reader should by now be on to this, but it is amazing how the intellectual leaders of our time lose sight of the critical distinction between the generative origin and its dependencies, between first cause and secondary causation.

Summary with some emphasis on neglected points

In this chapter we have applied the geometry of meaning to the problem of free will, using for this purpose Cassirer's

Determinism and Indeterminism in Modern Physics. The first part of my book has shown the necessity of four different types of knowledge (of which rationalism is but one). In this chapter we have seen how the deficiencies of objective reasoning get Cassirer into difficulties.

Reason, as we have shown, involves relating or comparing one thing with another, hence, *ratio* and *ratio*nalization. Reason is not direct knowing, but comparison of what has already been experienced. It cannot exist in a vacuum and it cannot deal with first cause.

Starting with the Laplacean intelligence that is able to know the exact position and motion of every particle in the universe, we find we cannot use this information without drawing on the principle of cause and effect, a principle which cannot be verified because it is not objective. It is a belief. We are thus forced either to deny the reality of cause and effect with Cassirer and others before him, or in accordance with our geometry to affirm a *nonobjective* reality which we call *projective.* We discover that Kant, in his plea for what he calls transcendental knowledge, does the equivalent.

This takes us to the notion of substance, whose nonobjectivity was to Berkeley sufficient reason for dropping it and depending only on ratio knowledge. At the same time, however, Berkeley refused to accept Newton's invention of the derivative, essentially a ratio of infinitesimals and devoid of substance.

This conflict between form, or what can be held in the mind as a concept, versus substance, or that which is felt emotionally, has received a great deal of attention and has been the cause of many divisions in philosophy.

We insist that both aspects are necessary—in fact, interdependent. In calling the former objective and the latter projective, we stress this interrelationship, which is like the interrelationship of direction in space.

Form and substance, however, are not opposite, but perpendicular (or complementary) in that they share the property of generality. Each has its own opposite. Form is opposed by the particular projective, and substance by the particular objective. This oppositeness has not been covered so far. Let us first consider the opposition of substance and the particular objective:

General Projective	*Particular Objective*
Belief	Fact
Substance	Object
Emotion	Sensation
Reality as felt	Reality as encountered

A difficulty is that we are inclined to think of substance as objective. Real, yes; objective, no. Objectivity has no self-evident reality. We cannot feel it. However, we *feel* a pain or an emotion, but we cannot communicate it. Objectivity is what we *can* communicate, what we can describe and define. The box is square, the table is round, etc.

On the other hand, properties such as solidity which are communicable because the experience of them is shared may seem objective, but they are not. Perhaps one of the reasons people talk about the weather—"Isn't it hot today?"—is that this involves a shared experience.

What is involved here is the difference between feelings and sensations. The two words are often used interchangeably, and I doubt that much can be gained by insisting on linguistic distinctions. The point is, rather, to uncover the basic dichotomy to which language refers.

Suppose we touch the coffeepot to see if it is hot. Within certain limits, our finger is an instrument which informs us of temperature, much as would a thermometer. This is sensation. If the pot is very hot, we may receive a burn, in which case we withdraw the finger quickly. We have experienced pain, a negative value which teaches us to avoid hot objects. This second reaction is of a more primitive type than is the informative reaction of sensation without pain. The sensation gives us objective information: whether the coffee is ready; the pain teaches us not to use fingers for extreme temperature measurement.

The sensation of smell is one that is so poorly developed in man that the word "smell" has become synonymous with value, "That deal has a bad smell," whereas with animals, smell is informative. It is a true sensation. A case is on record of a man whose sense of smell had been highly developed for a period of time. Later, when it had become average once more, he was asked about the difference between his present and former state. His answer was that when his sense of smell was acute, all smells were interesting.

But we must not lose sight of the forest for the trees. The overall principle is that there is an opposition between conviction of reality, which is projective and is essentially a belief, and the verifiable facts of the outside world, which are discovered only through work.

The other opposition, between form and function (or between general objective and particular projective), is even more difficult. The difficulty is that the former, which covers reason and concept, gets all the attention. It is the visible end of the stick. The other end is the knower himself, and cannot be objectified without ceasing to be itself. This is the predicament of providing scientific status for free will which, existing only in its activity, eludes identification and labels. In looking for it, we are in the predicament of Perseus who, in order to cut off the head of Medusa, had to look at her in a mirror lest he be turned to stone. Medusa is the intellect, which converts anything it deals with into objects (stone). To circumvent this, we must invert our approach (the mirror).

To the observer, free will is uncertainty; precisely because it is free, the reasoning mind cannot predict it and cannot relate it to something else; it is ir*ratio*nal, unrelatable.

Thus the rational mind has to acknowledge a higher principle, one that is indistinguishable from pure chance. So great is this hurdle that few philosophers can surmount it. The anonymous *Cloud of Unknowing,** written in about 1350, affirms this positive value of unknowing with a most beautiful simplicity that should be instructive to philosophers. Certainly, it is one of the most profound writings in the West.

Of course, our trump card is quantum physics, which has actually discovered, or been forced to recognize, that the quantum of action or quantum of uncertainty (for these are two names for the same thing) is a primordial ingredient of the reality of physics. But even here the rational mind refuses to accept first cause. It is not convinced by the creation of matter from light, for it says, "How can we say the photon is first cause when it itself has a cause? Does not light come from the sun?"

Here we should make a clarification. In order to say that A causes B, there must be in A that which implies B. Thus, if I put salt on the doorstep, it will melt the ice because there is a relationship between the melting point of ice and the salinity of water.

On the other hand, we cannot so relate cause and effect in the case where the pot boiling over "caused" the invention of

*A recent translation of *The Cloud of Unknowing* has been published, New York: Julian Press Inc. Also in paperback, New York: Delacorte Press (Delta Edition), 1974.

vulcanized rubber. Pots boiling over don't imply rubber tires, or even rubber.

The cause was Goodyear, whose intention to find a way of improving raw rubber had led him to cooking it on the stove. Now, it is true that there were other causes—hot stove, accident, etc.—but the first cause, Goodyear's intention, is the only one that *implies* improved rubber, and was responsible for his recognition of the solution when it accidentally occurred.

To fully account for the distinction between first cause and other sorts of causation, we need to recognize that generation itself has a cycle, birth, youth, maturity, old age, death, whose phases occur on four levels.

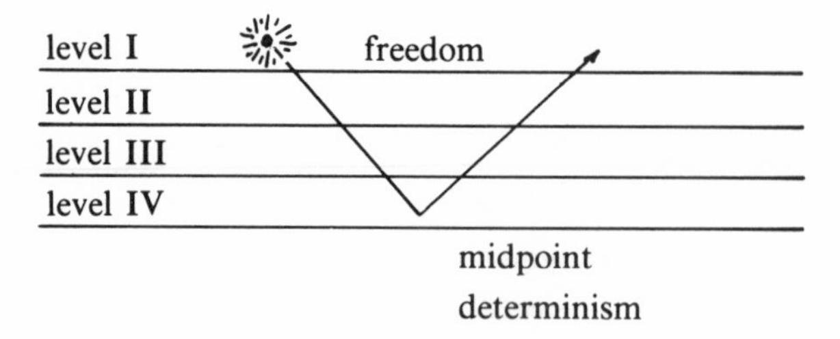

What is significant is that once generation is under way, development is determinate; causation, with the exception of the virgin birth at its midpoint, is based on antecedents.

At level I, however, generation is quite different. Here there are neither physical objects nor "laws" to move them. Previous events have no necessary influence on the future. It is similar to the situation before and after an explosion. The spark initiates a new generation whose development is not due to what has gone before. It is at this seed point that first cause operates. For example, a photon precipitates into an electron and a positron (pair creation), which dash off in opposite and unpredictable directions.

At this level too occurs the fertilization of ovum by sperm. While the DNA "blueprint" of each is definite, their combination, based on random pairing of genes, is unpredictable, a combination that never existed before.

Note, then, that first cause need not be without temporal antecedents. What makes it first cause is that these antecedents do not imply the result.

XIII Types of philosophy

The reader should be able to derive some benefit from the tool kit provided by the geometry of meaning. Geometry and the physical sciences are technical methodologies for control of physical objects. The geometry of meaning is also a technical methodology, but the objects it deals with are mental. By means of it we can keep our bearings in the face of the sophistry, evasions, and ambiguity which clutter the "educated" mentality.

"Mind is the slayer of the real," says the Zen Buddhist. "Therefore, we must slay the slayer." No, mind can serve us well, if it is made to do so and kept in its proper place.

How? By recognizing first that mind, even in the broadest sense, deals only with one aspect of the whole. It deals with *relationship;* it cannot and does not replace *action,* nor does it touch on that aspect of the whole that we are obliged, for lack of a better word, to call *states* (see beginning of Chapter II). Moreover, relationship is itself of four types. Only one or possibly two are properly called mental.

We have, in following Cassirer, witnessed the confusion to which reason can lead. Let us now attend to some of the great problems that philosophy exhibits, much as an art museum exhibits great paintings.

Earth versus air

This may also be expressed as sensation versus Intellect. As Will Durant puts it, "[the Greek Sophists said that knowledge] comes from the senses only . . . truth is what you taste, touch, smell, hear, see. What could be simpler? But Plato was not satisfied: if this is truth . . . the baboon, then, is the measure of truth equally with the sage . . . Plato was sure that reason was the rest of truth, the ideas of reason were to the reports of the senses what statesmen were to the populace, unifying centers of order for a chaotic mass."*

One would think Plato could have settled this issue once and for all. But then we have Locke, two thousand years later, credited with having discovered that knowledge comes from the senses.

From the point of view of the geometrization of meaning, or of the four elements, sensation supplies isolated instances (objective particulars) which intellect generalizes to create concepts (objective generalities).

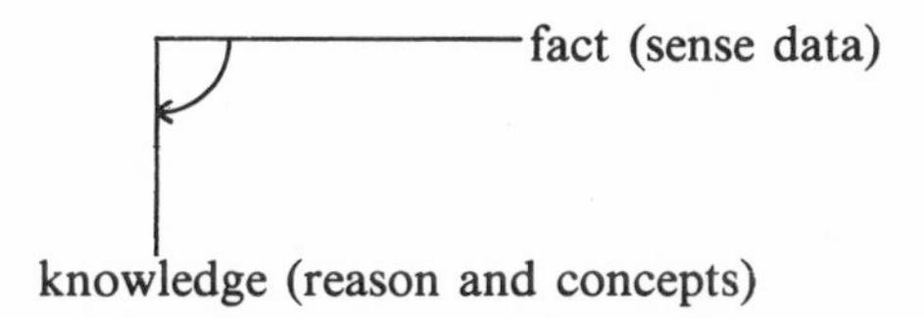

The shift from the category of facts or particular events to that of concepts is represented as a right angle, analogous to the shift from position (which we detect with the senses) to velocity (which we have to compute by observing a change in position and dividing by the elapsed time), a procedure that is carried out by the mind (the computer-type activity of mind, called intellect).

* *The Mansions of Philosophy* (p. 26). New York: Simon and Schuster, 1929.

This aspect of existence, the interrelationship of facts that can be mapped, computed, or communicated, is that available to the Laplacean intelligence. Its objectivity and computability give this aspect its special sanction in Western thought. But, as Cassirer said, "It is but a limited and partial aspect of the totality of being, of genuine reality."* (Despite this renunciation, we find Cassirer falling back into the trap when he later comes to uphold the findings of the theory or relativity, for relativity makes claims similar to those of the Laplacean intelligence in that it dismisses other kinds of knowledge.)

While the claims for objective knowledge are correct as far as they go, objectivity by itself is insufficient even for theoretical computation, which involves a projection from what is given to something else.

Water

Such projection is a category of knowing which takes many forms. We study the rhythm of the tides and "project" for the future. We study a company and project its future earnings, so that no matter how objective we are, we can make no use of this information without projection, that is to say, without assumptions. Assumption or belief is in a different category (water**) than the two categories we have considered so far (facts, or earth, and concepts, or air), a category we call general projective.

This is a difficult category largely because, unlike the general objective, it is neither communicable nor even definable (it is two right angles from fact).

**Ibid.*, p. 4.

**To resort to an altogether different kind of thinking, water symbolizes belief because there is a certain wetness about credulity; water suggests involvement, getting one's feet wet, being in the swim, etc.

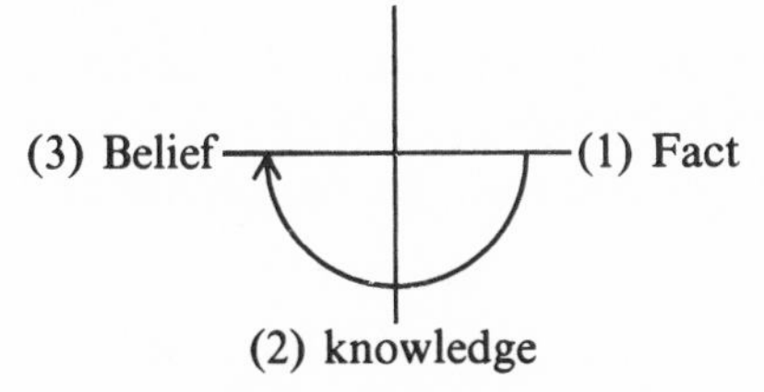

Despite its indefinability and its opposition to fact, we must recognize that it is the belief category that makes a thing *real.* We recall Hume's statement that causality is a "belief," and one that we are unable to give up. It is this very compulsiveness that carries the conviction of reality. We have already indicated that it is the basis for experience, to which it contributes feeling (as we *feel* acceleration, for instance).

This is the knowing which Kant distinguishes as prior to objective experience, calling it transcendental. His description of it as *a priori* confirms our insistence that it is ontologically prior to objective knowledge.

While Kant regards this transcendental knowledge as epistemological, we have been at pains to show that it is more than that; it is ontological. Thus energy and time exist, but are not objective. If you want to drive yourself crazy, try to think of time as objective. We say, "Give me time," but the transaction does not involve handing anything over. Likewise, energy is not objective in any sense of its being a definable object or having identity. Yet it is real, like pain, acceleration, fear.

The reader may find that he cannot go along with my statement that energy is not objective. "Surely," he will say, "the energy is 'out there' and therefore objective." This I would not dispute, if by "objective" you mean "out there." The designation "objective," however, is generally used in the sense that a statement is objective if it can be confirmed by other people; an entity is objective if it can be seen by more than one observer.

Is time of this nature? Can the present moment be

identified? If we say that when the bell strikes, it will be 10:00 P.M., we are identifying a moment in time, but when this moment occurs, it can never be experienced again. It cannot be summoned up for examination as one could a true object. The question of "out-there-ness" becomes meaningless about time. First it is in the future, suddenly it becomes the past. "It" has changed. What is the consistent part that would give it identity?

Energy, which conforms in every respect to substance, the *prima materia* of philosophers, also lacks the fixed character which would make identification possible. For example, it is not possible to say: "*This* kilowatt is moving through the wire and lights the book I read." When we measure a current of electricity on an ammeter, we are observing the effect of the current, not the current itself. But that, it might be claimed, is all we can expect.

Well, if that is all we can expect of objectivity, why can't I say my headache is objective because anyone can observe me holding my head and groaning? No, you do not feel the headache as I do. Similarly, the reality of electricity can be experienced if we take hold of the wire and get a shock. When we do this, it is not "out there," it is right here.

This is the special character of the nonobjective "level." We can say that it is out there, but to confirm the "out-there-ness," we must experience it right here, and we cannot erase the difference between this "reality" and "objectivity" by redefining words.

We are thus brought to a rather drastic revision of our view of the universe. We cannot insist that it be exclusively objective. The universe is also projective (or even subjective). This is not idealism or solipsism. The projectivity is not in me, it is in the universe. The universe is thinking or feeling itself into existence!

Fire

This takes us back to level I, the projective particular. We have placed "photons" here, by which we imply all electromagnetic radiation. This level, of course, has a correspondence to *purpose,* for this correlation was the basis of Chapter IX on purposive intelligence and the twofold operator.

Taking our cue from the interdependence of free will and purpose (purpose implies free will, and vice versa), we can put free will at this level. Since photons (at stage one) are quanta of indeterminacy, we thus have free will associated with indeterminacy.

What do we mean by indeterminacy? Indeterminacy is, of course, not a concrete thing or object. It is a bit like time in the expression, "Give me time." "Give me indeterminacy or give me death!" Death would be total determinism. "Dead men tell no tales." The opposite of death and determinism is life at liberty: indeterminism.

Let us return to Cassirer. His final rejection of indeterminacy (as equivalent to freedom) reads:

> "None of the great thinkers . . . ever yielded to the temptation to master them (the problems of free will) by simply denying the general causal principle and equating freedom with causelessness. Such an attempt is not to be found in Plato or Spinoza or Kant . . ."*

There are a number of ideas in this statement which we must untangle. In the first place, the thinkers mentioned did not know about quantum physics, and it is quantum physics that cuts through the double talk and forces the recognition of uncertainty.

Secondly, "by simply denying the general causal principle

**Ibid.*, p. 203.

and equating freedom with causelessness"—are we to understand that by this Cassirer means denying cause and effect? I see no need to deny cause and effect—unless by "the general causal principle" Cassirer means the edict against first cause implicit in determinism. But surely he does not mean that. There are many great thinkers who accept first cause. Even stuffy old Aristotle listed final cause as one of the four sorts of cause.

"And equating freedom with causelessness." I cannot see the relevance of this except as an indication that Cassirer is confusing the initial spark with the train of causation that follows in its wake.

Let us take another example. Suppose someone drops a lighted match on the dry leaves of a forest, and in short order there is a forest fire. We say the act was malicious or careless, and blame the miscreant. We would join with the moralist and condemn such freedom as "lawless behavior." I think this sentiment has intruded in Cassirer's analysis of the problem.

But there are several confusions here. While we condemn such action morally, we cannot deny that it is possible. Quantum theory, in fact, affirms it. Moreover, spontaneous combustion does occur, and if we could by act of law prevent all spontaneous occurrences, we would exclude all good acts too. In fact, the moral issue can exist only if there is this possibility of a first cause, the spontaneous act of kindness, the freely given smile. Cassirer knows this and we all know it, just as we know that Achilles *can* catch the tortoise. The problem is to answer the thought-stopping rationalism that says everything must have a cause. What causes first cause?

What quantum physics offers is testimony from the hard sciences that first cause *is*. It is everywhere (for example, mutation caused by cosmic rays). It is not so much that ethics must stoop to pick up the crumbs of the scientific banquet, as that the determinist must be made to do his sums properly and not misquote science.

The other confusion is that at no time has it been suggested, either by quantum theory or advocates of free will, that the laws of nature be set aside. No one is advocating causelessness. For it is precisely because free will occurs in an orderly universe that it can have far-reaching effects. If my will says "Forward!" and my legs are paralyzed, nothing happens. If the captain gives an order but the crew stages a mutiny, the captain's will is ineffectual. So we must see the problem not as free will versus determinism, but as free will *plus* determinism.

We may express this interrelationship somewhat paradoxically. An organism is effective insofar as it is both indeterminate as a whole and determinate in its parts. In other words, if all the parts are completely reliable and their actions determinate, the whole functions perfectly. If the whole functions perfectly and is operated by a free will, it cannot be predicted.

This is the answer to the computer stalemate, the problem of two equally matched opponents, both with perfect computers, A and B. Both have complete information and each can anticipate the other's moves. Now provide a random switch at the "top" of A and not at the other. A can anticipate B, but B cannot anticipate A. The random switch puts "life" into A. B remains an automaton, responding but not initiating.

Now, suppose B says, "Ah! I'll go A one better. I'll put two random switches!" What happens? The first switch goes into place at the "top," as with A, but the second must go somewhere else. This means that part of the organization separates out from the rest of the command. It becomes autonomous, and the net result is to weaken B.

Thus unity is very important to freedom of will. Without it we have divided action (civil war, mutiny, disease, paralysis). If we get millions of random switches, we get utter stasis: the vase on the mantle contains billions of molecules moving and vibrating, but the vase stands still.

This bears out a final point of great importance. The ideal freedom, or indeterminacy, which is freedom of the whole, is particular (and, of course, projective).

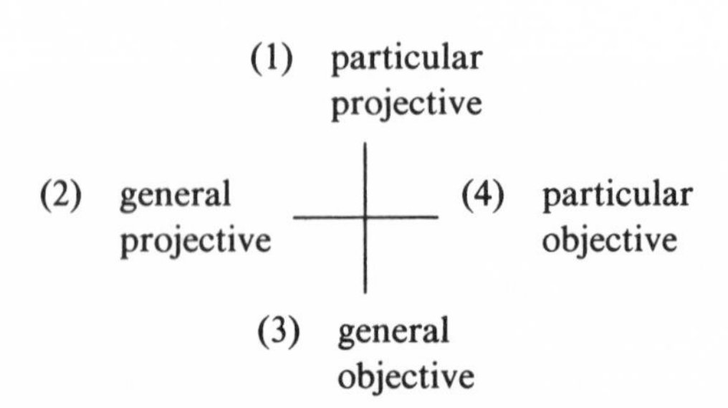

This rounds out the scheme. It shows that the particularity or uniqueness at 1 is indispensible. This characteristic of 1 emerges from other considerations. For example, it reflects the quantumicity, wholeness, or *oneness* of action. (Note that 1 and 4 deal with quanta. The quanta at 1 are quanta of action; the quanta at 4 are quanta of matter, molecules. The fact that atoms do not naturally occur in isolation may be significant. Molecules do occur singly, as in gases, or in ordered configurations, as in crystals, or singly again, as in DNA.)

Yet these technical considerations hide the special magic of this final category or aspect of action. To the ancients it was *fire,* but more in the sense of a spark that initiates; it is the creative, the *divine* spark, the spirit itself. This is higher mind, intuition. (Note again that intuition is particular; it guesses the exceptional; it does not work by rule, as does the lower or rational mind.)

Naturally, there is confusion about all these points. Cassirer, whose statement we have been trying to untangle, brings in intuition in the end:

> "Freedom (means) . . . determinability through the pure intuition of Ideas, determinability through a universal law of reason that at the same time is the highest law of being, determinability through the pure concept of duty in which autonomy, the will's self-ordering according to law, expresses itself; these are the basic criteria to which the problem of freedom is brought back."*

Intuition of ideas . . . universal law of reason . . . pure concept of duty . . . I respect Cassirer's attempt to express the inexpressible, but to me his statement would be improved if he omitted the "push me, pull you" evident in the conflict between freedom and conformity. I prefer to think of this higher type of "determinability" (and by this he means that free decision is effective, that is, determined) as directly opposite in nature to the predictable type of decision that activates the computer.

It responds to a "higher" law, or it responds to its own inner intuition of the future, but it does not react in conformity to the laws that govern and contain secular interchange. If it did, it would comply with the second law of thermodynamics (entropy), and join with molar matter and move lower and lower in the ladder of organization.

Résumé

We have now made our last circuit of the four types of knowledge. Let us enumerate some of the many forms the fourfold divisions of categories take, listing them in reverse order:

**Ibid.*, p. 203.

Categories:	**Fourth**	**Third**	**Second**	**First**
General Name	*Object*	*Form*	*Value*	*Purpose*
Relationships	From object to self	Within object	From self to object	Self's use of object
Aristotle's causes	Efficient	Formal	Material	Final
Four functions	Sensation	Intellect	Emotion	Intuition
Four measure formulae	Work	Power	Inertia	Action
Maps	Where you are	Map itself	Scale	Compass reading
Elements	Earth	Air	Water	Fire
Aspects of machine	Assembly	Blueprint	Material	Function
Actions	Control	Observation	Reaction	Spontaneous act
Stimuli	Fact	Concept	Belief	Guess
States	Establishment	Significance	Transfor-mation	Being

Below are tentative indications of how different philosophies tend to stress one (or two) elements at the expense of others [major emphasis = X; secondary emphasis = x].

	Object	*Form*	*Value*	*Purpose*
Empiricism	X			
Pragmatism	X		x	
Determinism	x	X		
Vitalism			x	X
Existentialism				X
Phenomenalism	X	X		x
Idealism		X		
Rationalism		X		
Logical Positionism	X	X		
Mysticism			X	x

Instinct

What is instinct? To call it an innate habit pattern isn't enough. We use the word to cover a rather wide range of phenomena, from the egg-laying habits of wasps to the intuitive hunches of humans that may cover the highest type of creative activity, as when we refer to mathematical intuition.

To start with something we can understand, let us take any ordinary action, such as walking, driving a car, speaking, writing, reading. All these activities have been learned by each of us with varying degrees of difficulty. In due course, however, we are able to walk, drive, speak, write, read "instinctively." Whether or not this is like the instinct of animals is not important. The point is that what was once practiced one step at a time and required great concentration becomes automatic and requires no attention; we drive without thinking, carrying on a conversation while we do so. What we have done is to go through the four steps of the cycle of action, until at 4 we learn the proper action which, *when it becomes habit,*

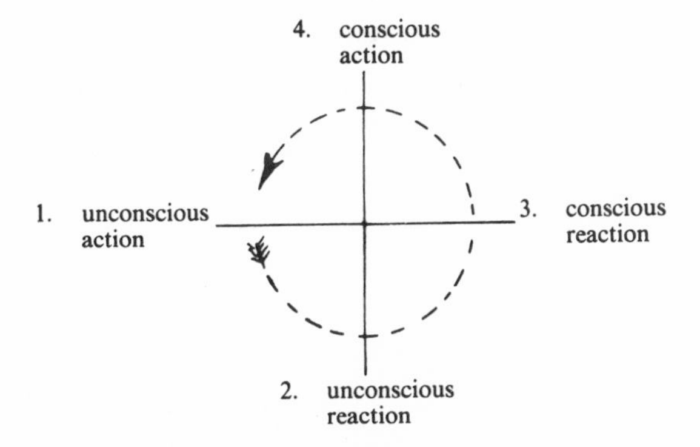

moves on to 1. Here 1 was the start and was entitled unconscious, and when it succeeds 4 it is conscious, but now it has learned something, and this knowledge becomes unconscious, or *instinctive.* (This is "fire," or *final* cause; note that it is final both in the sense that it is first, and in the sense that it is last.)

This is a very interesting thing. We have 1 as both naïve

and completely informed. Can you "play" the piano? Play chess? Play bridge? A child bangs his fist on the piano; he is playing on the piano. So then he is given piano lessons. He works hard at it. Perhaps ultimately he learns to "play" again, reeling off a Beethoven sonata like a concert pianist. Painting is similar. It is quite easy to turn out a passable sketch. Well, perhaps I'd better go no further, modern art, and all that.

In any case, there is a difference between the stages in the learning cycle, and the difference is greatest between the beginning and the end, even though both are intuitive. We are not forgetting that the cycle may go round and round. A competence which is learned provides the opportunity to undertake a cycle of greater scope. Thus we learn to read, then we read about something, say, integral calculus or quantum physics.

This is why instinct is both primitive and complex. This is why intuition is so inexplicable. This is why fire is the category that represents wholeness. As the origin of things, fire is wholeness, wholeness in the sense that an egg is a potential chicken. It is the origin, the seed.

But fire is also the whole when it is regained. It is the whole when the parts have been put back together. And this brings out why mind is "the slayer," as the Zen Buddhist puts it; why mind, which takes things apart, also creates the stasis, inaction, and yet it has to be gone through.

For mind is the midpoint of the cycle of action:

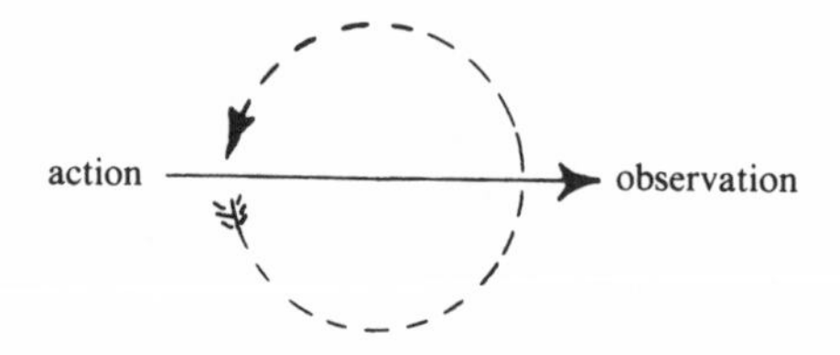

It stands exactly opposite the initial point of starting. It is the point of maximum distraction, separation.

But it would be a mistake to confuse what the mind is for what it does. Mind, though dispersed, *converges inwardly.* Its opposite, action or fire, though compacted, radiates outwardly. So we cannot pronounce judgment on the elements. We cannot say this is good, that bad. Universal process moves through them all and needs the contribution of each for its fulfillment. "We cannot imagine space extending only to the east," as the Chinese saying puts it.

Becoming

We can make a similar case for the importance of earth and water, for the importance of learning from practical experience, and for the importance of drawing on the deeper yearnings of the soul, whose messages, often in an unknown language, are signaled to us in dreams, in slips of the tongue, in mistaken assumptions by which we twist reality in ways that can be only our own projections.

So we stand, bubbles in the maelstrom of cosmic churning, starting boldly and innocently, receiving rebukes, becoming circumspect, learning carefully, until finally having learned one thing, we launch out again. Life consists of all phases of the cycle of becoming. All kinds of action, experience, and thought have their value. The optimum life does not imply sitting in the dead center, being nothing. It rather teaches that all things have their place and importance, and follow one another, like the rolling of a great wheel.

The Buddhists teach how to escape from the wheel, but I rather think they mean we should escape from endless repetition of the same cycle. Each new cycling should be a new adventure, incorporating, not repeating, what has gone before.

Where does it lead?

There is only one true escape from the wheel. This is to reverse our direction upon it. To exercise the option afforded by the twofold operator. In terms of the cycle of action, this reversal occurs at control point 4:

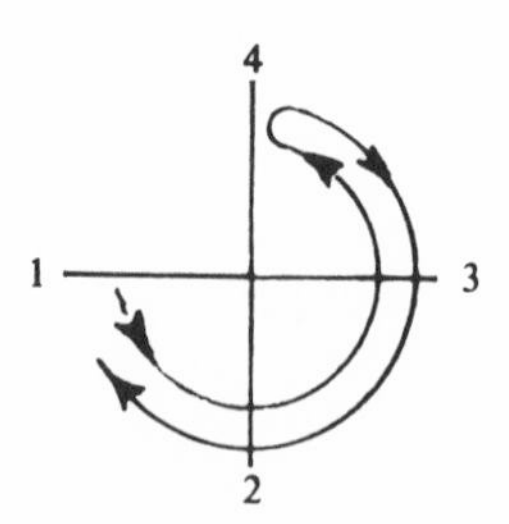

It may also be represented, as in Chapter IX, by an arc or V-shape:

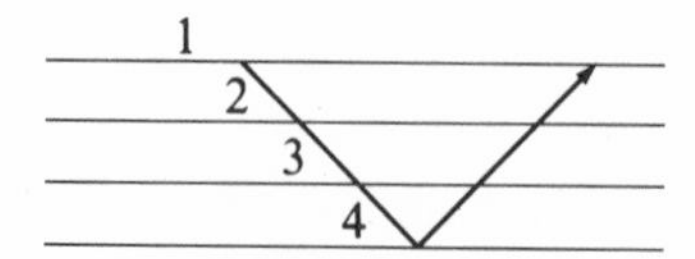

In either case, instead of moving counterclockwise from 4 to 1, which would precipitate it into a new involvement, the self makes use of what it has learned to extricate itself by mastering the laws of matter. It turns around and goes the other way.

In Chapter IX we indicated that the grand scheme of evolution followed this V-shaped arc from photons, through atoms, molecules, and cells, to animals and man. But each one of these stages itself involves a process of development, one that goes through stages (which we may call substages in order to avoid confusion with the major stages). Such is the case for man's evolution. According to the Hindus and, to some extent, the West (for example, Plato and Christianity before A.D. 553*), this evolution requires many lifetimes and, according to the scheme suggested by the geometry of meaning, it moves through four levels: first "down," then back "up," the change from down to up constituting man's "rebirth," or self-determined growth to higher status and eventual godhood.

Man's evolution and, for that matter, all evolution, follows, and is closely related to, the scheme of ontology represented in the four levels, but we cannot develop this subject here. Here we have been concerned with ontological and epistemological foundations. As in the foundation for a building, this has involved a great emphasis on the square, the compass, the level, and the plumb bob. It has been primarily a geometrical exercise.

In the last four chapters, we have engaged in a tentative and preliminary exercise by applying the scheme, set out in Chapter X, to human situations.

In Chapter XI, we indicated how steps can be taken toward correcting the piecemeal rationalism of Western thought by invoking the ancient concept of the four elements, followed by a demonstration of the correlation of the zodiac with what we call the Rosetta Stone: the arrangement of the twelve measure formulae of physics in a circular format.

In Chapter XII, we took up a typical philosophical problem, that of free will, using as a reference Cassirer's *Determinism and Indeterminism in Modern Physics.* Our endeavor here was, again, to show the use of square and compass to deal adequately with the ineffable and necessarily irrational aspect of existence.

In Chapter XIII, we went still further to indicate the scope

*When the belief in the preexistence of the soul was declared a heresy.

of the four elements, or aspects of totality, and their ability to interrelate different schools of philosophy. This chapter closes with the indication that despite the fixity of structural relationship, the circle of meaning is dynamic, the basis for an evolving process.

Not only must we recognize the variety of mutually contradictory qualities which go to make up the whole circle, but we must acknowledge two possible directions for moving around the circle. The first, or counterclockwise, direction, is the natural one and carries us deeper and deeper into involvement. The second, or clockwise, direction comes about when, having become involved, we are able to learn the law, put it in the service of the will, and thus evolve to a higher state.

The author

Arthur M. Young, inventor of the Bell helicopter, graduated in mathematics from Princeton University in 1923. In the thirties he engaged in private research on the helicopter, and in 1941 he assigned his patents to Bell Aircraft and he worked with Bell to develop the production prototype. In 1952 he set up the Foundation for the Study of Consciousness, which published the *Journal for the Study of Consciousness* and in 1972 the book *Consciousness and Reality,* edited by Arthur Young and Charles Musès. The Foundation has now been superseded by the Institute for the Study of Consciousness, located in Berkeley, California, and dedicated to the integration of scientific thought and the development of a science of mind/body interaction.

Index